FIELDWORK EXERCISES
Human and Physical Geography

M.H. Matthews and I.D.L. Foster

Hodder & Stoughton

LONDON SYDNEY AUCKLAND TORONTO

Acknowledgements

We would like to thank our colleagues at Coventry (Lanchester) Polytechnic for their valuable advice on many matters involved with these exercises. We are especially grateful to Brian Ilbery for his views on the human geography projects, particularly agriculture in the rural–urban fringe and urban retailing, and Michael Healey for his comments on intra-urban manufacturing change. Thanks are also due to Joy Summers and Joan James for their excellent typing. The idea for this book was not ours, Cherry Matthews continues to be a source of inspiration and Libby Foster has cast a constructively critical eye upon all our efforts.

The publishers would like to thank the following for permission to include copyright material:

Butterworth Scientific Ltd (London) for two illustrations from D Briggs: *Sediments*; Almqvist & Wiksell for a figure from 'Perception of Urban Retailing Facilities: an Analysis of Consumer Information Fields' by R B Potter (1979) Geogr. Ann. 61B 19–29; Edward Arnold (Publishers) Ltd for a figure from K J Gregory and D E Walling: *Drainage Basin Form and Process*; Her Majesty's Stationery Office for extracts from *Soil Survey Field Handbook* which are used by permission; Oxford University Press for a figure from R U Cooke & J C Doornkamp: *Geomorphology in Environmental Management*; the Royal Scottish Geographical Society for a table from 'The Assessment of Scenery of a Material Resource' D L Linton (1968) Scottish Geographical Magazine 84, 3 and D Herbert for Fig. 6.

© 1986 M. H. Matthews and I. D. L. Foster

First published in Great Britain in 1986
Fourth impression 1991

British Library Cataloguing in Publication Data

Matthews, M. H. (Michael Hugh)
   Fieldwork exercises in human and physical geography.
   1. Geography—Fieldwork
   I. Title   II. Foster, I. D. L.
   910'.724   G74.5

ISBN 0 7131 7357 2

All rights reserved. No part of this publication may be reproduced or transmitted in any form or by any means, electronic or mechanical, including photocopy, recording, or any information storage and retrieval system, without permission in writing from the publisher or under licence from the Copyright Licensing Agency Limited. Further details of such licences (for reprographic reproduction) may be obtained from the Copyright Licensing Agency Limited, of 90 Tottenham Court Road, London W1P 9HE.

Printed in Great Britain for the educational publishing division of Hodder and Stoughton Limited, Mill Road, Dunton Green, Sevenoaks, Kent by Athenaeum Press Ltd, Newcastle upon Tyne.

Preface

Fieldwork forms an important part of geographical learning. A number of 'A' level syllabuses already include compulsory fieldwork as an integral part of their assessment procedure; this may take the form of an examination paper which demands an awareness of field techniques or represent a project or set of assignments developed and executed by the candidate. Even if no formal component exists within a syllabus it can be argued that students derive considerable benefit from undertaking geographical inquiries of their own. This book attempts to meet these needs by providing a set of fieldwork exercises which could be carried out by individual sixth-formers.

The book is divided according to human and physical geography. Sets of exercises are provided in each section. These have been chosen so as to be of relevance to most 'A' level syllabuses and are designed to introduce students to a wide range of data sources, various techniques of data collection and different methods of data analysis. The projects are set in different environments, for example, urban and rural, upland and lowland, but most can be carried out regardless of pupils' home location.

Each exercise has been set out in a systematic and structured manner. The topic has been briefly introduced and its aims clearly stated. Guidelines have been given in terms of how students should go about collecting the data and of what pitfalls and problems they should be especially aware. Ideas about how analysis could be carried out and relevant techniques designed to assist interpretation, both qualitative and quantitative, have been suggested. In conclusion, a brief list of references have been offered in order to place each project within a broader context.

Discussion within the book is brief but pertinent. The idea is to provide students with a series of frameworks and suggestions about suitable 'A' level projects, set out in a step-by-step format. Each exercise is of sufficient scope and flexibility to suit the expertise and interests of most 'A' level students.

Additionally, there is a brief section which attempts to introduce students to questionnaire design, sampling and some of the problems associated with carrying out physical geography projects in the field.

The book is aimed at the individual researcher and the projects presume only one person will be involved in data collection. However, most of the exercises can easily be extended for group work and teacher use.

Contents

1	An introduction to fieldwork techniques	6
2	Accessibility and service provisions in rural areas	9
3	Landscape evaluation	13
4	Tourist pressure upon the countryside	16
5	The internal scructure of the city	19
6	Social and environmental quality: an intra-urban investigation	23
7	Studies of the Central Business District	27
8	Urban neighbourhoods	30
9	Urban retailing: a behavioural investigation	34
10	Intra-urban industrial change	37
11	Agricultural land use in the rural–urban fringe	40
12	Service hierarchy and spheres of influence	42
13	Routine meteorological observations and microclimate studies	47
14	Forest interception studies	51
15	Measurement of the water balance in soil profiles	55
16	The measurement of river velocity and river discharge	57
17	Morphological, geomorphological and slope mapping	61
18	Analysis of hillslope erosional processes	66
19	Human impact on river channels	68
20	Analysis of the physical properties of sediments	73
21	Describing and measuring soil properties in the field	78
22	Analysis of soil moisture content	85
23	An analysis of plant productivity and growth habit in bog communities	88
24	The analysis of vegetation associations	91
Index		96

1 An introduction to fieldwork techniques

Planning a questionnaire survey

Many of the human geography projects suggested in this book rely upon primary data collection by means of questionnaire surveys. The following notes attempt to provide some guidelines for the design and organisation of these surveys.

1 When planning a project do not be over-ambitious. Avoid the temptation of covering everything that seems interesting. Collect only that data which will be used.

2 Target population. Define the population that will be interviewed (respondents), e.g. adults, chief wage earner of household or any particular sub-section.

3 Sampling. Decide upon the method of sampling. The object of this exercise is to gain a representative group of the defined population. Three sampling techniques are commonly employed, random, systematic and stratified and these are discussed in the next section.

The standard procedure for selecting a random sample of adults is to use the Register of Electors which provides the names and addresses of every person registered to vote, generally listed in street order. These lists are located in large local libraries and with the local authority. The alternative is to select a particular household but be clear about which individual is of interest. Resist the temptation just to interview the person who answers the door.

4 Sample size. For most projects it will be impossible to carry out a large survey, but generalising about 5 or 10 respondents is very dangerous as answers are likely to be atypical and unrepresentative. As a rule of thumb, aim for about 20 to 30 cases and this should be enough to permit some statistical analysis.

5 Questionnaire design. Designing a questionnaire involves backward planning. Ideally, set up a series of hypotheses/sub-hypotheses linked to the project's aims. Questions should be related to each of these hypotheses. There is no right or wrong length for a questionnaire, but avoid the temptation to collect data for its own sake. Each question should have a particular purpose.

6 Types of data. The questionnaire will probably include different types of data: background or classificatory data which relates to the socio-personal characteristics of the sampled population, e.g. age, sex, length of residence etc; activity data, informing on such matters as what people do, how often and when; attitudinal data, relating to what people think about certain issues.

7 Format. Considerable care should be given to the organisation of the questionnaire. Avoid asking all background questions at the beginning or lumping these altogether later on. In the first place attempt to gain the respondent's attention. Start with some easy lead-in questions and build-up to more complex issues. Background information should be introduced at different times.

8 Types of question. There are two types of question: open questions, which allow respondents to answer freely in their own words, such as, 'where do you usually shop for groceries?'; closed questions, where answers are predetermined by a set of categories, such as 'how long have you lived in this village?'; (i) less than 1 year (ii) 1–5 years (iii) more than 5 years. The advantage of the former is that respondents are not influenced by any determining factors, but this may cause difficulty when making generalisations and problems may be experienced when attempting to note down people's verbal replies. With the latter, care must be taken to select a sufficiently wide set of alternatives from which people can choose, although analyses should be much easier, especially if the answers are numbered to permit the frequency of set replies to be assessed.

9 Working. Keep the questions short and simple. Avoid all jargon and assume nothing. Try to ensure that the questions can only be interpreted in one way. Phrase questions so that they do not lead the respondent to give a particular answer. With attitude questions, where respondents are asked to indicate agreement or disagreement with a question, it is often useful to employ a graded scale. For example, rather than a 'Yes' or 'No' answer consider whether a five point scale would introduce much needed variation, such as, (i) strongly agree (ii) agree (iii) not sure (iv) disagree (v) strongly disagree. Bear in mind how the results are to be analysed as this will influence the way in which questions are presented. For example statistical description and testing is often aided if answers can be grouped into categories enabling frequency values to be derived.

10 Pilot survey. Even with the most carefully planned questionnaire there are bound to be some problems which have not been anticipated. In which

case it is always useful to carry out a pilot survey. Interview only about five people. This should be enough to highlight any inconsistencies or major flaws in the structure or the wording of the questionnaire.

11 Conducting the interview. Door to door interviews can seem intimidating, but they offer a very successful means of gaining primary data. At all times appear confident and well organised; slip-shod presentation will only result in a high number of refusals or careless answers. Before undertaking the survey inform the respondent who you are and the purpose of the investigation. Above all, stress the confidential nature of the inquiry.

Measurement and sampling

1 The physical geography sections of this book rely heavily upon basic measurements conducted in an objective sampling framework. It is important that all measurements are performed with great care since statistical analyses of the data and interpretation of the results will only be as good as the basic data permit.

2 A variety of levels of measurement can be made, and the level should be selected carefully because this to a large extent determines the type of statistical test which can be performed on the data. Four levels are commonly recognised; from lowest to highest these are:
i) Nominal scale – a classification based on mutually exclusive categories but not arranged in any hierarchy (e.g. land-use types; woodland, grassland, market gardening, cereal crops).
ii) Ordinal scale – this scale of measurement is based on the ranking of observations but no account is made of the difference in magnitude between the ranked observations (e.g. the classification of slopes as steep, moderate or gentle, or of soils as acid, neutral and alkaline).
iii) Interval scale – at this scale, observations are not only ordered but the size of the difference separating any two observations along the scale is variable. The interval scale, however, is not proportional; i.e. a temperature of $20°C$ is not twice as hot as $10°C$ because the zero value at this scale is set at an arbitrary point (the freezing point of water).
iv) Ratio scale – this is the highest level of measurement and relates to the existence of an absolute zero in the level of measurement. Measurements of length, distance, rainfall depth etc. fall into this category because, for example, the differences between 5 and 10 mm of rainfall is 5 mm and the difference between 20 and 25 mm of rainfall is 5 mm. The ratio of the differences, unlike temperature, are also the same if we use imperial units of measurement rather than the metric scale (this is not true when we convert $°F$ to $°C$ and vice versa).

3 The choice of statistical test is also important. For most of this book, non-parametric statistical tests are used. These are advantageous because:
i) They do not assume a normal frequency distribution for the data.
ii) They may be used with data measured on the ordinal and nominal scale.
iii) They are easy to compute with a pocket calculator.
iv) They are useful with small sample sizes.

On occasions, it has been suggested that the mean, standard deviation and coefficient of variation are calculated. It should be noted that these calculations assume a normal distribution for the data.

4 Careful selection of a sample is essential for many of the projects, and two types of sampling are usually identified:
i) Purposive or subjective sampling
ii) Objective sampling

Purposive samples are subjectively chosen by the researcher, the most common type being the 'case study', such as choosing a drainage basin for detailed analysis of hydrological processes or fluvial deposits. Several problems arise from the use of purposive samples.
i) Many samples of this type are not typical but are 'ideal' samples, are likely to be biased and may not provide truly representative results.
ii) It is possible for two independent researchers to select different samples and, not surprisingly, arrive at different conclusions on the basis of the analysis.

Probability samples are selected randomly in such a way that the researcher cannot influence which samples are chosen to represent the population and which are not. In physical geography, we may be interested in either:
i) sampling from a defined area (e.g. from a river bed).
ii) sampling along a linear feature (e.g. from the top to the bottom of a slope).

Since measurement of all the pebbles in a stream bed, for example, could take a very long time, we need to take a 'representative sample' which represents the characteristics of the whole population.

Two methods may be used for areal sampling.
i) Define the area to be sampled by means of a sample grid (usually two tape measures laid out at right angles to each other) to give a 'Northing Scale' and an 'Easting Scale'.
ii) Use a table of random numbers to provide co-ordinates for a 'Northing' and an 'Easting'.

iii) Locate the point of intersection of these co-ordinates at right angles to the grid, which determines the point at which the sample is taken.

Alternatively, a systematic sample may be taken.
i) Define the area to be sampled as given above, randomly locating the origin of the grid.
ii) Collect samples at the points of intersection at pre-determined distances along the 'Northing' and 'Easting' axes (e.g. every 0.5 or 1 m).

This method assumes that the population is randomly distributed and that no bias is introduced by using a regular sampling scheme.

For linear features, a transect is used to collect the basic information. A transect is a pre-determined line (usually laid out with a 30 or 50 m tape measure), with the origin of the transect randomly located. Samples are taken at regular intervals along the transect (e.g. at 1 m, 5 m, or 10 m intervals), until the appropriate sample size has been reached.

5   It is important that the criteria of any statistical test are met, such that the sample obtained is free from bias and is precise in terms of defining the characteristics of the population from which the samples are drawn. Where there is a choice of sampling scheme, the one which makes the sample easiest to collect should be used provided that this does not cast doubt on the validity of the sample.

A note of caution

Fieldwork can be academically rewarding, enjoyable and perfectly safe provided adequate precautions are taken. The following notes cover only the more important points.
i) Always wear suitable clothing; warm and waterproof rather than fashionable. Stout waterproof footwear is also recommended.
ii) When working in upland areas or remote locations, take a large scale map, a compass, spare waterproof clothing and a supply of food and drinking water. (Do not sample the moorland streams — they are often contaminated.) Before carrying out the project, find out where you are on the map and take a compass bearing to the nearest road or settlement.
iii) When working in remote locations, tell the local park rangers or police where you are and when you expect to return. Do not go out on your own, and take a basic first aid kit.
iv) All land belongs to someone, it is impolite not to ask permission for access to carry out field experiments (and will stop access for others in the future). Remember that some areas are used for military purposes. These are clearly marked on maps and in the field. Avoid these areas for fieldwork.
v) Treat the environment with respect. If small soil pits are dug, refill them carefully and leave the area as you found it. Large holes in the environment are a hazard to animals and walkers alike.

References

Dixon, C.J. and B. Leach (1978) Questionnaires and interviews in geographical research. *CATMOG* no. 18. (Geobooks) Norwich.

Gardiner, V. and R. Dackombe (1983) *Geomorphological Field Manual* (Allen and Unwin) London.

Goudie, A. (ed) (1981) *Geomorphological Techniques* (Allen and Unwin) London.

Hanwell, J.D. and M.D. Newson (1973) *Techniques in Physical Geography* (Macmillan) London.

Hoinville, G. and R. Jowell and associates (1977) *Survey Research Practice* (Heinemann) London.

Moser, C.A. and G. Kalton (1971) *Survey methods in social investigation* (Heinemann) London.

2 Accessibility and service provision in rural areas

1. Introduction

An important trend within rural Britain has been the tendency for service outlets to become fewer and more widely spaced into larger centres. Problems of accessibility seem an inevitable consequence.

Accessibility is a term which is difficult to define and measure. It refers to people's ability to reach things which are important to them. The reasons why someone or something may be inaccessible or difficult to get at may be quite varied: i) access problems may arise because of the distribution of services in relation to likely users — spatial issues; ii) inaccessibility may be more evident for certain social groups of people, such as the elderly or the low-paid, whose circumstances may hold back their ability to travel to reach non-local services — structural issues; iii) different groups of people in society may have different appreciations of access, with some poorly served groups not aware of their relative disadvantage, and others, who may not be so badly off, complaining bitterly about local difficulties — awareness issues.

2. Aims

The problems of accessibility can be studied within different environments, but the project outlined below relates to investigations which could be carried out in a rural area.

Geographers have been interested in studying rural accessibility at two scales:
i) inter-village variations in accessibility, to provide an index of the *spatial* pattern of accessibility.
ii) intra-village variations in personal accessibility, to examine variations in accessibility between social groups, hence focusing upon *structural* and *awareness* issues.

This project looks at all three aspects of accessibility although it would be possible to focus upon just one of these in any study.

3. Method

3.1 Study area

Ideally the study area should contain a range of villages of different size, some located near an urban centre, others more distant and perhaps, located away from main roads. Bear in mind that you will need to travel around this area in order to collect the data. An alternative would be to make a comparison of a few villages of contrasting character.

It would be useful to find out the population size of each village, by reference to the parish listings in the 1981 UK Census. Where villages are very small a parish may include more than one village.

3.2 Spatial issues

In order to see whether accessibility varies from place to place an index needs to be developed which measures this component. The one suggested below (Table 1) employs a point-score technique to record the presence or absence of selected services. An essential part of this index is a weighting system which scores the availability of certain facilities more highly than others. For example, is access to a chemist worthy of the same score as access to food-shops? — how important are bus services to local movement? — and more generally, which services should be included in this survey?

3.3 Structural and awareness issues

These issues can be investigated by means of a questionnaire survey of residents (Table 2). It may be impossible to carry out a survey within each village, so the total index scores derived above could be used to select places with different access characteristics.

The questionnaire should aim to highlight differences in personal accessibility. In addition residents' awareness of their accessibility problems need to be sought. To ensure that the attitudes of different social groups are expressed, sampling could be carried out within different parts of the village. For example, interview people living in (i) owner occupied, (ii) council and (iii) private rented sectors, and this should ensure some social class variation. Before carrying out the survey be clear who it is you want to interview within a household. Remember that answers will vary according to whether that person is at home or at work during the day.

4. Analysis

4.1 Spatial issues

4.1.1 Calculate the total index score for each village. Record the values onto a base map. Use these figures to draw an isopleth map of accessibility. Which areas or groups of villages are characterised by

Table 1 An index of accessibility and service provision

Part 1	Public bus accessibility	Points
1.	*Travel to work*	
	i) bus arriving in a town before 8 am, daily	2
	ii) bus arriving in a town before 9 am, daily	1
	iii) number of towns served by these buses	
	> 5	3
	2–5	2
	1	1
2.	*Travel to entertainment*	
	i) a bus leaving a town after 9.30 pm, daily	2
	ii) a bus leaving a town after 9.30 pm, at least once per week	1
3.	*Travel to shopping*	
	i) for each possible return journey to a town per week, giving at least 2 hours stay there and leaving and returning to the village between 8 am and 5.30 pm.	1
4.	*Sunday service*	
	i) > 5 buses	3
	ii) 2–5 buses	2
	iii) 1 bus	1
5.	*Length of journey to high order towns* (high order, needs to be defined in the context of the study area)	
	i) < 30 mins	3
	ii) 30–60 mins	2
	iii) > 60 mins	1
Part 2	**Access to fixed services**	
6.	*Educational facilities*	
	i) secondary school in village	3
	ii) primary school in village	3
	iii) state school within 8 km of village, on bus route	2
	iv) state school within 8–16 km of village on bus route	1
7.	*G.P. surgery*	
	i) in the village	3
	ii) within 8 km of village, on bus route	2
	iii) within 8–16 km of village, on bus route	1
	Access to chemist, general foodstore, post office, entertainment, can be scored same as above.	
Part 3	**Access to mobile services**	
8.	*At least one visit per week:*	
	i) Doctor's surgery	2
	ii) Library	1
	iii) District nurse	1
	iv) Food vans	1
	v) Private bus/mini-bus service	1
	TOTAL INDEX SCORE	____

high or low values? Does accessibility seem to relate to such variables as population size of the village, distance from an urban centre, distance from main roads, or the surrounding physical environment? What conclusions can you draw from your map about levels of accessibility in the study area?

4.1.2 The total index score comprises of three main parts: public bus accessibility; access to fixed services; access to mobile services. For selected villages draw divided bar charts based on this information (Fig. 1). Do different villages experience varying accessibility problems? If so, where are these villages located?

4.2 Structural and awareness issues

In order to see whether different social groups experience or perceive different accessibility problems it is necessary to categorise the respondents to the questionnaire according to which social characteristics are of most interest e.g. age, social class (see Chapter 5), mothers with young children.

4.2.1 Enter the results for each social group upon a separate master data sheet: the columns should

Table 2 Questionnaire on accessibility

1	Sex of respondent i) male ii) female	
2	Length of residence in the village? i) < 1 year ii) 1–5 years iii) 6–10 years iv) > 10 years	
3	How many cars or other vehicles are at the disposal of the household?	
4	Do you have a driving licence? i) Yes ii) No	
5	If yes, do you have access to vehicle? i) at all times for most purposes ii) restricted access during day iii) restricted access during shifts iv) never normally available to me	
6	Do you use public transport? i) Yes ii) No	
7	If yes, how frequently do you use it? i) daily ii) > twice a week iii) twice a week iv) once a week v) once a fortnight vi) < once a fortnight vii) other	
8	How satisfied are you with the bus service? i) very satisfied ii) satisfied iii) dissatisfied iv) very dissatisfied v) don't know	
9	Are there any problems with the bus service? i) frequency ii) cost iii) routes iv) reliability v) other	
10	Where do you usually go for the following services and how do you get there? Location Transport e.g. i) Baker ii) Post Office iii) Chemist iv) Doctor v) Primary/Secondary School vi) Cinema	
11	Do you feel that you have any difficulties in getting to these services? i) Yes ii) No	
12	If yes, please specify which and why?	
13	Age group of respondents? i) < 15 ii) 15–30 iii) 31–45 iv) 46–60 v) > 60	
14	Number of children aged? i) < 5 ii) 5–10 iii) 11–16	
15	Occupation of chief wage earner in household?	
16	Place of work of chief wage earner in household?	

represent the answers to the questions, the rows relate to each person interviewed. The frequency response to a question, converted to a percentage, provides the basis for a comparison of results. For example, do people from social classes 1, 2, 3 experience or perceive different accessibility problems to those from social classes 4, 5, 6? Do mothers with young children at home during the day encounter different accessibility problems to those people in employment?

4.2.2 Whether different social groups encounter or perceive different accessibility problems can be

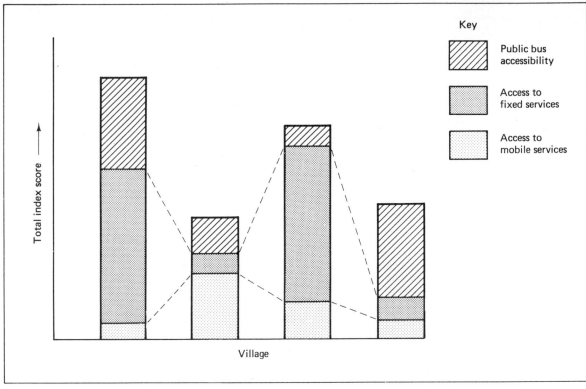

Fig. 1 Components of accessibility

statistically examined with a chi-squared two-sample test. For the formula, the rows would consist of the two social groups under investigation, for example, mothers with children and single people; social classes 1, 2, 3 and social classes 4, 5, 6; over 60 and retired and under 30. The columns relate to the behaviour or attitude which is being considered and the different categories allowable in the answer. For example, there are seven categories to question 7 and five categories for question 8. Enter only the frequency values into the cells of the contingency table. Remember that it is important to discuss the findings of your statistical tests. If a significant difference exists between the two populations, consider why this may be. On the other hand, if no difference is suggested by the results, then this is an important conclusion and should not be dismissed. In order to assist explanation go back to the contingency table and look at the distribution of expected and observed frequencies, where large differences are evident then this indicates deviations away from what was predicted.

4.2.3 Flow maps of where different social groups travel for services and work provide useful supplementary information.

References

Cherry, G (1976) *Rural Planning problems* (Leonard Hill).

Moseley, M.J (1979) *Accessibility: the rural challenge* (Methuen).

Phillips, D and Williams, A (1985) *Rural social geography* (Blackwell).

3 Landscape evaluation

1. Introduction

The term 'landscape' is used here to refer to the natural environment. A number of trends have contributed to a growing awareness that the landscape is an increasingly vulnerable resource. First, urban pressure upon the countryside is growing, the product of continual urban expansion and increasing recreational demand by urban inhabitants. Secondly, many rural areas are threatened by the competing demands made by such land-uses as forestry, water, extractive industries and military establishments. Thirdly, recent agricultural developments have led to considerable changes within the countryside often modifying or altering the traditional rural scene.

An aspect of the landscape which is of considerable importance is its scenic quality. Information of this kind is important to resource managers and planners concerned with such issues as landscape preservation, conservation, improvement as well as recreational policy. A fundamental problem is how to measure visual appeal. Two general categories of study have approached such assessment:
i) those which focus upon observable components of landscape;
ii) those which assess personal response to scenery.

2. Aims

The project aims to develop an awareness of the uses, limitations and problems of landscape evaluation projects by examining two assessment procedures:
i) Linton's component technique — whereby the separate elements that make up the landscape are identified and measured.
ii) Personal technique — whereby an individual's reactions to an environment are expressed and measured.

3. Method

3.1 Study area

The project outlined below is intended to be carried out in areas where the natural environment predominates, although these techniques can be modified to incorporate urban places. For greatest interest an area suffering from the effects of competing land-uses or threatened by urban pressure would provide an ideal location in which to undertake this investigation. Another strategy would be to compare the results derived from an area of protected land, such as an Area of Outstanding Natural Beauty with those nearby and not safeguarded by legislation.

The area should be of sufficient size to include a variety of landscape components and yet be accessible to an individual researcher. The shape of the study area need not be regular and may be influenced by an issue of local concern, for example, a linear coastal stretch of protected and undesignated land, or a section of pressurised land such as the Vale of Belvoir.

3.2 Sampling framework

It is important to superimpose upon the study area a grid within which data will be collected. A useful framework is provided by the 1 km grid found on OS maps. Different approaches can be adopted; data can be gathered at specific points or generalised for areas. For example, for each ¼ km grid square an overall assessment of the landscape can be made and a value derived for the whole area or at a particular location the complete view can be visually surveyed and a value obtained to reflect that point. Alternatively, a linear transect can be taken, whereby a route across the study area is followed. Whatever method is preferred it is usual to follow a systematic procedure, so that a regular sequence of locations are sampled. When determining the sampling interval attention needs to be given to the landscape under investigation. For example, a finer collecting frame is needed where local variation is great and where views are impeded by such landscape elements as narrow valleys, mountain ranges and forests, whereas a broader interval can be used in areas of relative uniformity.

3.3 Linton's technique

This is an attempt to measure landscape quality by focusing upon two principal aspects of the environment, land form and land-use. Each of these is assessed separately by comparison to a given set of categories (Table 3). A total point score is derived for each sampled location, whether part of a grid square, or specific point.

Table 3 Landscape assessment

```
1.   Landform type
i    Mountains (mean relief ≥ 610m) +8
ii   Bold Hills (mean relief ≥ 366m) +6
iii  Hill Country (mean relief less than 305m) +5
iv   Plateau uplands (mean relief less than 91m) +3
v    Low uplands +2
vi   Lowland 0
Note that some of this information may be map-derived.
2.   Land-use
i    Wild Landscapes +6
ii   Richly varied farming landscapes +5
iii  Varied forest and moorland with some farms +4
iv   Moorland +3
v    Treeless farmland +1
vi   Continuous forest −1
vii  Urbanised and industrialised landscapes −5
Total the point scores and add bonus points as:—
3.   Bonus points
i    For areas with water in the foreground and middle
     distance of the view +2
```

This technique works best in areas of wildscape and highland Britain. For lowland Britain and many coastal areas an alternative set of categories and points will need to be developed. In these instances try to think of features of the landscape which add to its visual quality ('attractors') or have a negative scenic value ('detractors'). Remember that these facets will be influenced by your own perception of 'good' and 'bad'.

3.4 Personal technique

This method attempts to measure an individual's reactions to a landscape. The first step is to decide upon a set of adjectives and their opposites which you consider can be used to describe your own preferences towards a landscape. One way to do this would be to take three photographs of different parts of the study area and to consider how each view differs from the other. The adjectives derived in this way can be arranged upon a grid and matched by an opposite, in order to broaden the range of possible responses (Table 4). In this process make sure that all positive and negative adjectives are grouped separately. Refer to Roget's Thesaurus or a dictionary to help you choose opposite adjectives. A point system, providing scoring options between +3 to −3, can then be introduced to help scale individual judgements of an area. Where an adjective is considered irrelevant to a landscape a zero score would be awarded. The grid is now complete providing a means by which personal responses to a landscape can be measured. It is possible to adapt this technique for group use by ensuring that the choice of adjectives reflects group preferences.

Table 4 Personal adjectives for environmental assessment

LANDSCAPE EVALUATION (−3 to +3)		VIEW		
		1	2	3
Boring	Interesting			
Dull	Exciting			
Ugly	Beautiful			
Filthy	Clean			
Noisy	Quiet			
Unattractive	Attractive			
Smooth	Rugged			
Straight	Indented			
Spoilt	Untouched			
Functional	Historic			
Avoid	Go to			
Inaccessible	Accessible			
Ordinary	Spectacular			
Depressing	Stimulating			
Dislike	Like			
Common	Unusual			
Locally admired	Widely admired			
Displeasing rural	Pleasing rural			
Low	Elevated			
Enclosed	Open			
Monotonous	Various			
Neglected	Well maintained			
Polluted	Unpolluted			
Personal	General			
Go in passing	Visit regularly			
Displeasing man-made	Pleasing man-made			

For every location the whole view should be assessed and a total score derived by adding together the values given to each opposite set of adjectives e.g.

View 1. boring/interesting +2
         filthy/clean   −1
         quiet/noisy   −2
         etc   ——
         Total score   +1

4. Analysis

The analysis should consider the variation of landscape quality within the study area and its implications to local land-management policy, as well as providing a comparison of the two assessment techniques.

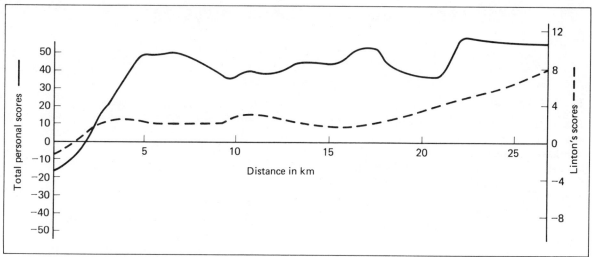

Fig. 2 Variations in landscape quality

4.1 Patterns of landscape quality

4.1.1 For each technique: if the whole area within each grid cell has been assessed draw a choropleth map of landscape quality within the study area. The range of values needs to be determined and a suitable scale chosen for the shading categories. Where point values have been derived draw an isopleth map of landscape quality. If the study area is linear in form, such as a stretch of coastline, produce a graph of total score (y axes) against distance (x axes) (Fig. 2). Provide an overall description of landscape quality. Consider the incidence of high and low scores and examine their relationship to issues of local concern, such as landscapes threatened by development or protected by legislation. What implications do these findings have to land-management policy within the area?

4.2 Comparison of the techniques

Determine the extent to which the scores obtained by each technique are related to each other. On the basis of this information consider the relative merits of each technique for assessing landscape quality.

4.2.1 Plot a scattergraph of Linton's scores (y axes) against the personal grid scores (x axes). Draw a best-fit line by eye through the points. The scattergraph provides a visual assessment of the relationship between the scores. By reference to the best-fit line, values which sharply deviate from the overall trend can be isolated and examined. Why have these occurred? What features of the landscape led to these discrepancies?

4.2.2 Measure the strength of this relationship by using Spearman's rank correlation coefficient and test for significance by using the Student's t-test. For each view rank the scores from high to low e.g.

	Linton's score	Rank	Personal grid score	Rank
View 1	29	1	32	2
2	8	3	−6	3
3	28	2	40	1

Comment upon the derived coefficient and examine the implications of this result. If a high positive correlation is evident this suggests that the two techniques provide similar assessments of landscape quality. On the other hand, if the relationship is a weak one or even not significant this indicates inconsistencies between the scoring procedures. This may result from the failure of one or both techniques to describe the landscape in an adequate manner.

4.2.3 For each adjective used on the personal grid plot a graph recording the total positive and negative values under each heading. Do the same for Linton's technique, in this case plotting only the total values. What do these results suggest about the usefulness of the different variables employed by each technique. For example, are some adjectives more useful than others? Which adjectives score highly, which are associated with negative scores and which are seldom used? Consider Linton's assessors in the same way.

References

Coppock, J.T and Duffield, B.S (1975) *Recreation in the countryside: a spatial analysis* (Macmillan) London.

Linton, D.L (1968) 'The assessment of scenery as a natural resource', *Scottish Geographical Magazine*, 84, 3, 219–38.

4 Tourist pressure upon the countryside

1. Introduction

The continued growth of car ownership coupled with improved road communication and increasing leisure time has meant that more and more people are seeking outdoor recreational activities away from their homes. Such mobility enables people to visit interesting and beautiful environments, especially during weekends. The parts of Britain which have been most affected by outdoor leisure pursuits are the Areas of Outstanding Natural Beauty and Great Landscape Beauty, National Parks, Country Parks, and those tracts of wildspace which are close to the largest urban centres. The scope of these pressures are likely to increase: without effective monitoring and carefully planned rural management strategies considerable damage to the countryside seems an inevitable consequence.

2. Aims

2.1 First, this project sets out to examine some of the recreational characteristics of a rural area experiencing considerable weekend pressure from tourists. Attention will focus upon three issues: the characteristics of the tourists, the activities that they carry out and the movement of visitors from outside and within the designated area.

2.2 Secondly, by considering the distribution and demands of land-uses within the study area an attempt should be made to devise a land-management scheme which would balance these conflicting interests.

The accompanying notes relate to a project designed to be undertaken within a National Park, although the ideas and methods can easily be modified to relate to other rural environments.

3. Method

The data collection falls into two parts: the first focuses upon tourists and is based largely upon questionnaire mapping; the second, relies upon land-use mapping and the questioning of people resident within the study area.

3.1 Questionnaire mapping for tourists

An interesting technique for gaining information upon tourists, their activities and movements is questionnaire mapping. A suggested format is provided in Fig. 3. The advantage of this technique is that impressions can be gained about overall flows and pressure points without the need to visit many different sites. The questionnaire is aimed at the car driver on arrival at various sampling places within the study area. The questions included in Fig. 3 are intended as suggestions and there is ample scope to extend the questionnaire. For example:

i Mark with dotted lines your intended route beyond this stopping place
ii Mark with ? where you intend to stop along this route
iii How often do you come to this area?
   a) > than once a week
   b) once a week
   c) once a fortnight ... etc.

The places where people go for recreation will be influenced by their knowledge of the local area. Often tourist pressure-points build-up because visitors remain unaware of alternative, yet similar locations nearby. The level of information that people have about different places in the locality may be assessed in two ways:

i Ask respondents to state, 'what are the following places noted for?'
ii Provide a series of photographs taken within the study area and ask people to recognise their whereabouts. In both cases every right answer scores a point.

It may by worthwhile seeking variation by carrying out the questionnaire at different kinds of site within the study area. For example, a riverside site could be compared with a village centre or an area of high physical amenity.

Useful supplementary information on the pulse of tourist flows can be gained by undertaking car counts regularly during the day. On the hour count the number of cars parked at a site.

The extent to which a place reaches its 'physical capacity' can also be assessed at these times. This is the actual capacity of a site in absolute terms: for example, the number of parking places in a car park in relation to the number of cars parked.

3.2 Land-use mapping and questionnaire for residents

Within the study area it is likely that many different interests are represented, these may include farming,

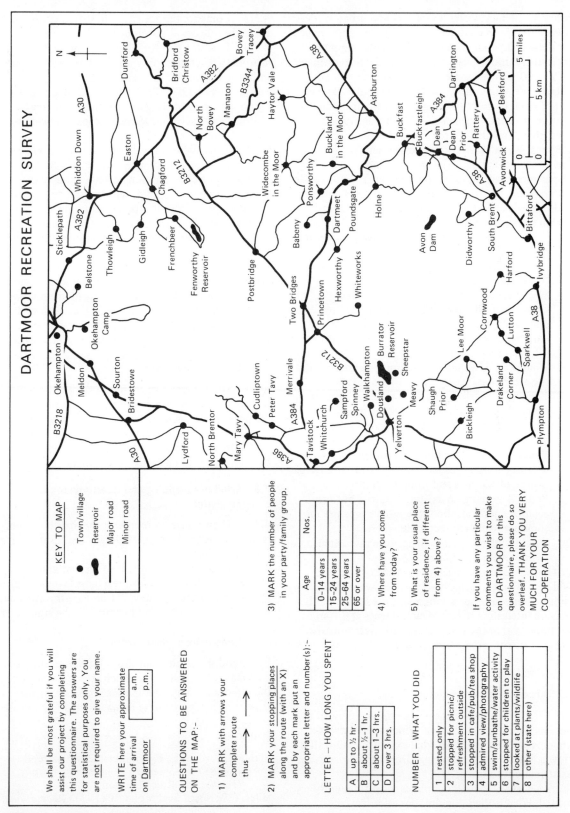

Fig. 3 Questionnaire map

forestry, military, water, extractive and urban land-uses. Attempt to consider the varying demands made by these competing land-uses by broadly mapping their distribution. The Second Land Use Survey maps (1:25,000) provide some insight into their overall pattern, although, these data are now often 20 years out-of-date. In order to bring this information up-to-date contact those agencies which manage local land-uses: these may include the Forestry Commission, Regional Water Authorities, Department of Environment (extractive), County Planning Departments (agriculture and settlement), Countryside Commission, Sports Councils and Tourists Boards. In addition, seek out the attitudes of those resident within the study area. For example:

i) Do you think that the number of tourists visiting the area is?:
   a) too great
   b) at right level
   c) too few
ii) Grade the following facilities within the area on a scale of 1 (very poor) to 5 (very good); car parking, picnic areas, road signs, toilet facilities, information centres.
iii) What policies would you like implemented within the area?
   Policy 1 — no change to what exists at present
   Policy 2 — stricter control of tourists — reduce the size of car parks — limit the number of picnic sites and stopping places.
   Policy 3 — increased provision for tourists — enlarge car parks, widen roads, provide additional toilets, picnic sites and stopping places.
   Policy 4 — maintain existing facilities but use signposts and guidebooks to relieve pressure on some areas and redirect it elsewhere.
iv) What problems and advantages are associated with tourism in this area?

On this occasion a balanced sample of residents will be difficult to achieve so these answers will simply represent impressions about the local area. It may be interesting to interview residents at different locations within the study area: some near tourist pressure-points, others further away.

4. Analysis

4.1 Tourist characteristics

4.1.1   Produce flow maps of the routes followed by visitors within the study area. Which road networks are most or least used? To what extent is road traffic flow a function of signposting? Do particular sites act as attractions to traffic?

4.1.2   List and rank those places most frequently visited. What is the mean duration of a visit?

4.1.3   Compare the activities most frequently undertaken by tourists at different sites.

4.1.4   Examine the distances travelled that day by visitors to reach the study area. Categorise the results into distance zones: for example, 0–7.9; 8–15.9; 16–23.9; 24–49.9; 50–99.9; 100–200 km. Draw choropleth maps distinguishing the percentage of visitors from each zone. Similar maps can be drawn for each site.

4.1.5   Those people travelling less than 25 km can be termed local visitors, whereas those travelling over 50 km can be classed as long-distance tourists. Compare these two groups with respect to the routes followed, places visited, trip structure, length of stay and patterns of activity. Are the recreational requirements of the long-distance tourists the same as for more local visitors?

4.1.6   Do long-distance tourists appear less informed about the area, visiting only the better known most accessible places? Determine the mean informational level scores for short- and long-distance visitors. Is there any relationship between the distance visitors live from the study area and their knowledge of it?

4.2 Site characteristics

4.2.1   Draw a graph to represent the total numbers of cars parked on each hour at a site. At what times are peak flows reached? Are there any periods when 'physical capacity' is reached or approached? If more than one site has been visited compare these results. Do some sites fill up earlier than others?

4.3 Land-use mapping and questionnaire for residents

4.3.1   Draw a map to represent the distribution of land-uses within the area. Highlight those sites of high tourist pressure.

4.3.2   Examine the attitudes of residents towards tourists. Which strategies do residents favour in terms of land-management policies? Compare the attitudes of residents from different locations in the study area.

4.4 Land-management planning

By careful selection of all this information suggest a land-management strategy which attempts to reconcile the interests of these competing land-uses, at the same time recognising the demands made by tourists.

References

Blunden, J, Haggett, P, Hamnett, C and Sarre, P (1978) *Fundamentals of human geography* (Harper and Row) London.

Miles, C.W.N and Seabrooke, W (1977) *Recreational land management* (Spon) London.

5 The internal structure of the city

1. Introduction

There have been many attempts to describe the internal structure of western cities (Fig. 4). These models characterise urban land-use arrangements as zones, sectors or nuclei. The extent to which a city conforms to these land-use patterns provides a good basis for project work.

2. Aims

The focus of this investigation is deliberately restricted. First, the method only relies upon secondary data sources, with the intention of demonstrating some of the value of the modern census. Secondly, rather than attempting a full summary of all aspects of urban form only a part of its structure will be examined.

3. Method

3.1 Data sources

Details about the residential structure of the British city can be acquired from a variety of secondary data sources, principally the census and local rate assessments.

3.1.1 The modern census

The censuses for 1961, 1966, 1971 and 1981 provide the most comprehensive sources of information upon the internal structure of urban areas. Even though the census collects a wide range of information about individual people, it is not concerned with people as individuals. The census is taken solely to compile

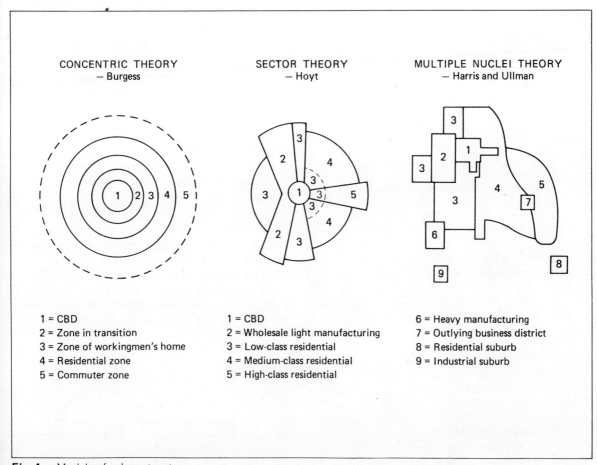

Fig. 4 Models of urban structure

statistics about groups and categories and, in particular, geographical areas. Small Area Statistics (SAS) are provided at two scales; the enumeration district (ED) which is the smallest area, and the ward, which consists of a group of enumeration districts. Each ED consists of about 150 households or 500 people, whereas wards can vary between 8000 to 30 000 in size. Information is also available for National Grid Squares in England and Wales, although at a much higher cost. Maps of the boundaries of EDs and wards are produced on 1:10,000 Ordnance Survey Maps. The map sheets cover an area 5 × 5 kilometres, that is four sheets for each 10 kilometre National Grid Square. Copies of these and a full explanation of the tables and terms used in the 1981 census (SAS) are available from:

England and Wales:

Census Customer Services
OPCS
Titchfield
FAREHAM
Hants. PO15 5RR

Scotland:

Census Customer Services
GRO (Scotland)
Ladywell House
Ladywell Road
EDINBURGH EH12 7TF

Table 5 Registrar General's socio-economic groups (From: Census of Occupation)

1	Employers and managers in central and local government, industry, commerce etc. — large establishments (with 25 or more employees)
2	Employers and managers in central and local government, industry, commerce etc. — small establishments (with 25 or fewer employees)
3	Professional workers — self-employed
4	Professional workers — employed
5	Intermediate non-manual workers
6	Junior non-manual workers
7	Personal service workers
8	Foremen and supervisors — manual
9	Skilled manual workers
10	Semi-skilled manual workers
11	Unskilled manual workers
12	Own-account workers (other than professional)
13	Farmers, employers or managers
14	Farmers, own account
15	Agricultural workers
16	Members of the armed forces
17	Indefinite (inadequately stated occupations)

Dependent upon the size of the city, the following information should be extracted from the Census for each ED or Ward.
(i) Social class areas: By plotting the distribution of selected social classes in a city, patterns may emerge which can be compared to those described by the various land-use models. Contemporary SAS do not provide a direct commentary upon class but information is collected on occupations. Since 1951 occupations have been classified into socio-economic groups, each of which is supposed to contain people whose social, cultural and recreational standards of behaviour are similar. The categories used in the 1981 census are shown in Table 5. These socio-economic groups can be collapsed into socio-economic classes.

Table 6 Socio-economic group and equivalent class (From: General Household Survey (1975)

Socio-economic class	Socio-economic group	Description
1	3, 4	Professional
2	1, 2, 13	Employers & managers
3	5, 6	Intermediate & junior non-manual
4	8, 9, 12, 14	Skilled manual (plus non-professional own account workers)
5	7, 10, 15	Semi-skilled manual & personal service
6	11	Unskilled manual

In practice it is usual to combine the first three and last three of these classes:

Social class	Socio-economic group
Non-manual workers, i.e. middle class	1, 2, 3, 4, 5, 6, 13
Manual workers, i.e. working class	7, 8, 9, 10, 11, 12, 14, 15

For each ED/Ward calculate the percentage of these two social classes resident in the area.
(ii) Housing quality: Variations in housing quality within each ED/Ward can be measured by considering the occurrence of variables which measure amenity, such as percentage of households lacking basic amenities (hot water, bath, inside w.c.); the percentage of households lacking exclusive use of a bath and inside w.c.; percentage of households experiencing overcrowding, (i.e. households with more than 1.5 persons per room).
(iii) Residential areas: Groups of variables can be used in combination in order to gain an impression of different residential areas. For example, the census offers data upon tenure, distinguishing between owner occupation, council tenancy and different

forms of rented accommodation. Although the percentage of each of these in an ED/Ward gives some insight into internal structure, a more interesting picture emerges if social class is related to these housing categories. For example, the distribution of owner occupation alone says little about the social composition of the population found within those areas. However, if social class is considered as well then those owner occupied areas associated with the residence of high or low classes can be assessed. Many pairs of variables can be considered in this way, e.g. class and other tenure groups; household amenities and tenure; ethnicity (percentage of residents born in New Commonwealth) and tenure — all of which contribute to a better understanding of urban structure.

3.1.2 Rateable values

These provide important supplementary information upon housing quality and provide an indirect commentary upon the social class composition of residents. Unlike the census, information is available for every residential property and so the areal patterns derived from this source are not restricted by an ED/Ward framework. Rateable values are available for inspection within local authority rates departments.

Derive the mean rateable values of properties along selected streets. It may be convenient to sample properties when undertaking this assessment. Alternatively, superimpose a grid over the study area with its origin in the extreme south-west corner. The size of the grid is dependent on map scale and convenience. At the mid-point of each cell determine the rateable value of five properties (two on either side plus the selected property). The mean value for these properties should be entered as a 'spot height' at this point.

4. Analysis

In all the cases outlined below the mapped distributions should be compared to those anticipated by the urban land-use models. Special attention should be given to those factors which may lead to deviations from these models.

4.1 Social class

Draw choropleth maps to represent the percentage of high and low social classes within each ED/Ward. Map those ED's/Wards with above average levels of high and low social classes. Clearer patterns may emerge with the use of location quotients. For each ED/Ward determine the location quotients for particular social classes and produce choropleth maps of the resulting values. The location quotient shows the extent to which each of a set of areas departs from some norm and so is a measure of concentration. For example, location quotients can be calculated to reflect the distribution of low social classes. In this case, the location quotient relates the proportion of lower social class households to total household population in an ED to the proportion of lower social class households to total household population in a city.

$$Q = \frac{\text{Number of lower social class households in an ED}}{\text{Number of households in an ED}} \div \frac{\text{Number of lower social class households in the city}}{\text{Number of households in the city}}$$

A location quotient of 1.0 shows the ED has the same proportion of lower social class households as the city as a whole and the higher the Q value the greater the degree of concentration within an ED.

4.2 Housing quality

Choropleth maps can be drawn for each variable but more interesting patterns will emerge if housing quality indicators are ranked and mapped. The 'worst' five ED's/Wards on each indicator can then be plotted and their location examined. A composite map to reveal the areas of poorest housing quality can be drawn by adding together the rank for each variable within an ED/Ward and mapping the 'worst' five districts. Conversely, maps of good housing quality can be produced in a similar manner.

4.3 Residential areas

Draw a graph, one axes of which should be labelled percentage owner-occupation in an ED/Ward, the other expressed the percentage of working class in an ED/Ward (Fig. 5). Each area within the city can be located in this grid and a shading scale devised to distinguish its four principal quadrants. This information can be used on the base map to distinguish different types of residential area.

4.4 Rateable values

For each street determine the mean rateable value. Using a suitable graded scale draw a map to highlight variations in rateable values.

If a grid system has been used draw an isopleth map of rateable values, using the mean values as 'spot heights'.

This information should be used in conjunction with that derived from the census to highlight variations in housing and residential structure.

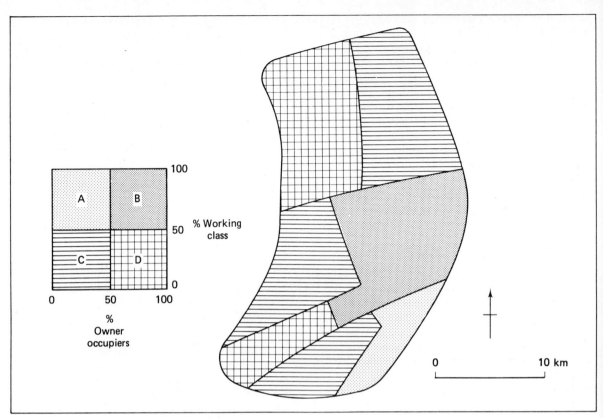

Fig. 5 Residential areas: a classification

References

Carter, H (1982) *The Study of Urban Geography*, (Edward Arnold) London.
Herbert, D.T and Thomas, C.J (1982) *Urban geography: a first approach*, (Wiley) London.

6 Social and environmental quality: an intra-urban investigation

1. Introduction

This project provides an introduction into the geography of social well-being. Such studies are concerned with comparing areas and their populations with respect to local issues which influence the quality of people's lives. Commonly two approaches have been used to make such assessments. The first attempts to measure the quality of the local environment by using a pre-arranged scoring system ('objective' approach). The other method focuses upon the attitudes and opinions of the people who live in the area ('subjective' approach). In this project both approaches are combined, providing a chance to compare 'objective' and 'subjective' assessments of a local area.

2. Aims

The project aims to examine the extent to which variations exist within an urban area in terms of (i) environmental quality, and (ii) perceived quality of life.

The first aim is explored by using 'objective' measures, whereas the second relies upon 'subjective' assessments by local residents.

3. Method

3.1 An assessment of environmental quality

The classification represented in Table 9 is adapted from schedules developed by the Nottinghamshire County Planning Department for use in assessing environmental quality in residential areas. Penalty points are awarded to each location according to the relative levels of deficiency noted in specific areas. The maximum number of penalty points which can be given to any one location is 70. A useful feature of the classifications is that this total is divided between four distinctive elements of the environment: 'appearance' (20 points), 'access' (20 points), 'amenity' (15 points), and 'provision' (15 points). This allows the variable nature of environmental deficiencies within towns to be identified. Inner areas of large cities, for instance, are likely to suffer badly from appearance and low amenity, but may be partially compensated by favourable access characteristics.

3.2 An assessment of perceived quality of life

People's sense of satisfaction with their lives and their overall sense of well-being can be assessed by looking at seven important aspects of life (Table 7). Interviewees are asked to respond to each of these aspects by allocating a score within the set scale.

Table 7 Perceived quality of life

	Points
Job	0 – 5
House	0 – 5
Public transport	0 – 5
Primary education	0 – 5
Secondary education	0 – 5
Health services	0 – 5
District	0 – 5
Total score	

Scale
0 Not relevant
1 Extremely dissatisfied
2 Dissatisfied
3 Satisfied
4 Highly satisfied
5 Extremely satisfied

Satisfactions with aspects of the local district as a place to live can also be measured in a similar way

Table 8 Satisfaction with aspects of the local district as a place to live

	Points
Services	0 – 5
Shops	0 – 5
Peace and quiet	0 – 5
Appearance and tidiness	0 – 5
Opportunities for entertainment	0 – 5
Lack of crime	0 – 5
Personal security and safety	0 – 5
Neighbours	0 – 5
Total score	

People's sense of satisfaction may change over time. Table 10 seeks to explore how people feel about their overall quality of life over time, employing the prescribed scale.

Table 9 Classification for assessing environmental quality

	Element *Appearance*	Penalty Points
Intrusion of non-conforming uses in residential areas e.g. industry	A Exclusively residential uses fully separated from other use zones B Limited infiltration of non-conforming uses C Substantial infiltration of non-conforming uses	0 1–3 4–5
Landscaping/visual Quality	A Mature, good quality trees; well placed and well-kempt grassed spaces B Insufficient poor quality trees; poorly placed, and/or unkempt grassed spaces C Total, or almost total, lack of trees/grassed spaces	0 1–3 4–5
Note: The incidence and visual quality of gardens is considered separately.		
Townscape/Visual quality	A Attractive built-environment B Some drabness within the built-environment C Excessive drabness within the built-environment	0 1–3 4–5
Appearance of gardens/yard	A Predominance of tidy/well screened gardens and/or yards within the study zone B Some intrusion of unkempt/poorly screened gardens and/or yards C Predominance of unkempt/poorly screened gardens and/or yards	0 1–3 4–5
	Element Appearance Score	____
	Access	
Access to Primary School	A Primary school within 5 minutes walking distance (550 metres) and involving no main road crossing(s) B Primary school within 5 minutes walking distance but involving main road crossing(s) C Primary school 5–10 minutes walking distance but involving no main road crossing(s) D Primary school more than 10 minutes walking distance but involving main road crossing(s) E Primary school more than 10 minutes walking distance (1100 metres)	0 1 2 4 5
Access to other facilities (shops, pub, doctor)	A Shops, public house and doctor all within 5 minutes walking distance B Shops and doctor within 5 minutes walking distance C Shops and public house within 5 minutes walking D Shops only within 5 minutes walking distance E Public house and doctor within 5 minutes walking distance F Doctor only within 5 minutes walking distance G Public house only within 5 minutes walking distance H No facilities within 5 minutes walking distance	0 1 1 2 3 4 4 5
Access to park/public open space (P.O.S.)	A Park/P.O.S. within 5 minutes walking distance and involving no main road crossing(s) B Park/P.O.S. within 5 minutes walking distance but involving main road crossing(s) C Park/P.O.S. 5–10 minutes walking distance but involving no main road crossing(s) D Park/P.O.S. 5–10 minutes walking distance but involving main road crossing(s) E No park/P.O.S. within 10 minutes walking distance	0 1 2 4 5
Access to public transporation	A Public transport route within 5 minutes walking distance B Public transport route 5–10 minutes walking distance C No public transport route within 10 minutes walking distance	0 2 5
	Access Score	____
	Amenity	
Traffic	A Full separation of pedestrian and normal residential traffic B Limited intrusion by through traffic/some intrusion of traffic of unsuitable character C Substantial intrusion by through traffic of unsuitable character	0 1–3 4–5
Noise	A Acceptable residential standards, i.e. normal speech possible B Slightly above acceptable residential standard, i.e. limited speech interference C Above acceptable residential standard, i.e. normal speech difficult at some times	0 1–3 4–5
Air pollution	A Negligible (or non-existent) B Light C Heavy	0 1–3 4–5
	Amenity Score	____

		Provision	
Garaging/Parking provision	A	Full provision of garaging/parking facilities	0
	B	75%–95% provision of garaging/parking facilities, i.e. limited on-street parking	1
	C	50%–74% provision of garaging/parking facilities, i.e. some on-street parking	2
	D	25%–49% provision of garaging/parking facilities, i.e. substantial on-street parking	3
	E	0%–24% provision of garaging/parking facilities, i.e. excessive on-street parking	5
Garden provision	A	Full provision of adequate gardens or communal/incidental open space: all requirements satisfied	0
	B	Insufficient provision of adequate gardens, or communal/incidental open space	1–3
	C	Excessive lack of gardens or communal/incidental open space	4–5
Provision of neighbourhood amenities, such as street lighting, telephone kiosks, post-boxes and bus shelters.	A	Full provision of all neighbourhood amenities within five minutes walking distance	0
	B	Insufficient provision of neighbourhood amenities, i.e. some amenities absent, within five minutes walking distance	1–3
	C	Total or almost total lack of all neighbourhood amenities, within five minutes walking distance	4–5
		Provision Score	
		TOTAL ENVIRONMENTAL SCORE	

Table 10 Overall sense of satisfaction with quality of life

	Points
Now	0 – 5
Five years ago	0 – 5
Five years ahead	0 – 5

3.3 Data collection

This investigation focuses upon variations within an urban environment. Two strategies are suggested.

3.3.1 Strategy 1

Superimpose a grid of convenient dimensions upon the study area. At the mid-point of each cell (or as near as possible) provide an assessment of environmental quality employing the proposed classification. Using this information it should be possible to choose areas of contrasting scores within which to carry out the questionnaire surveys. For example, dependent on the size of the urban area, total scores could be used to distinguish environments of different quality, or the results derived from each part of the index, that is, 'appearance', 'access', 'amenity' and 'provision', could offer a means for selecting areas of varying character. Within these selected areas a house-to-house survey should be undertaken in order to gain an impression of perceived quality of life. Ideally the sample will consist of residents who are inhabitants of the commonest type of house found locally. For example, if an area is predominently of council housing with some owner occupation the survey would focus upon the former.

3.3.2 Strategy 2

If the urban area is particularly large and diverse it would be better to concentrate upon selected districts from the outset. Choose areas known to be of contrasting character, for example, an area of owner occupation and a council house development, or an inner city site and suburban location. Within these areas both methods of assessment would be employed. Decide upon a sampling framework for data collection. This may consist of a street by street assessment or a fine grid could be superimposed upon the study area, the mid-points of which defining the sites for the schedules to be carried out.

4. Analysis

4.1 Environmental quality

The purpose is to examine whether variation exists and of what kind between areas in terms of environmental quality. If grid-squares have provided the collecting frame it should be possible to produce choropleth maps of environmental quality. The data should be grouped into classes and colour coded. Five separate maps can be drawn, recording total scores and 'appearance', 'access', 'amenity' and 'provision' scores. Where a small number of areas are being compared, description would be aided by divided bar charts, representative of both the total penalty score and its components.

4.2 Perceived quality of life

Compare the results derived from each area.
4.2.1 Determine the mean scores for —
 a) each aspect of life
 b) levels of satisfaction with each aspect of the local district
 c) overall quality of life, now, then and in the future

Do people respond positively (i.e. values > 2.5) or negatively (values < 2.5) to these different aspects of their life and environment.

4.2.2 The collective attitudes of people from different areas towards their perceived quality of life should be compared. Establish whether the results from these areas are similar or are different from each other by comparing the medians of the samples. People's levels of satisfaction with each aspect of life can be examined in this way.

4.3 Relationship between 'objective' and 'subjective' assessments of well-being

4.3.1 Examine the extent to which variations in environmental quality are matched by variations in perceived quality of life. Consider the total scores derived by the two techniques. For example, do residents of those areas with high penalty points on the environmental scale perceive a low level of well-being or are these two conditions unrelated?

4.3.2. Consider the implications of these results, especially towards providing a strategy for monitoring the quality of life. In so doing examine whether 'objective' measures of well-being offer a meaningful insight into the way in which people perceive their quality of life?

References

Coates, B.E, Johnson, R.J and Knox, P.L (1977) *Geography and inequality*. (Oxford University Press) Oxford.

Knox, P.L (1975) *Social well-being: a spatial perspective*. (Oxford University Press) Oxford.

7 Studies of the Central Business District

1. Introduction

An important part of the western city is the Central Business District (CBD). Typically, this area is made up of concentrations of retailing, consumer services, public and commercial office activities and an array of cultural, entertainment and hotel activities. Land-use in this part of the city is usually the most intensive. Within the CBD distinct areas associated with particular business activities can be identified. The land-use arrangements are constantly evolving: in time both the internal geography of the CBD and its overall extent reveal patterns of change.

2. Aims

2.1 To attempt a delimitation of a CBD using a variety of techniques.
2.2 To examine the internal structure of a CBD.
2.3 To consider the changing nature of a CBD over time.

3. Method

3.1 Delimitation of the CBD

3.1.1 Central business index

This technique was developed for American cities with their distinctive block structure but may be applied to the British city.

Stage 1: Map and classify all the land-uses in an area assumed to be the CBD into central and non-central types. Non central uses include permanent residence, government and public buildings, organisational establishments (churches, colleges, etc.), wholesaling, commercial storage, warehousing, industry (except newspapers), vacant sites and buildings. All other functions can be considered central uses.

Stage 2: For each city block, or area separated by street intersections, pace out the ground floor area devoted to central land-uses and make an assessment of the upper floor area given over to the same usage.

Stage 3: Calculate the central business height index (CBHI) of each block:

$$CBHI = \frac{\text{Total floor area devoted to CBD uses}}{\text{Total ground floor area}}$$

This represents the number of floors associated with central land-uses in the block if these uses were spread evenly over it.

Stage 4: Calculate the central business intensity index (CBII) of each block:

$$CBII = \frac{\text{Total floor area devoted to CBD uses}}{\text{Total floor area}} \times 100$$

This expresses the total floor area devoted to CBD uses in percent within a block. To be considered as part of the CBD a block should have a CBHI value of at least 1.0 and a CBII value of at least 50 per cent.

3.1.2 Rate index

Land values will be high within a CBD and will decline towards its edge. A number of studies have suggested that there is within a CBD a peak land value intersection (PLVI), often a major road junction, around which land values are arranged. Gross rateable value provides a means of representing land-value. Such data are available for public inspection at the rates office of the local authority. In order to take into account variations of floor space occupied by central properties calculate a rate index which can be used to distinguish the limits of the CBD:

$$RI = \frac{\text{Gross rateable value of a property}}{\text{Ground floor area}}$$

It may be more convenient to sample central properties rather than attempting a full inventory; for example, a systematic sample whereby every third property is assessed.

3.1.3 Core-frame concept

Another method by which the CBD could be differentiated is by careful land-use mapping. The 'core' (central CBD) is defined by:
i)   intensive use of land, with considerable evidence of multi-storey buildings;
ii) concentrations of retailing, consumer services, offices, hotels and entertainment facilities;
iii) congestion;
iv) high levels of pedestrian flow;
v)   dominance of intra-urban communication linkages;
vi) little residential population.

The 'frame' (peripheral parts of the CBD) is characterised by:
i)   less intensive use of land and less vertical development of buildings;

ii) a mixture of functions, such as wholesaling, light manufacturing, car sales and service, transport termini, specialised professional services and car parks;
iii) lower levels of pedestrian flow and increased traffic movement;
iv) dominance of inter-urban communication linkages;
v) some inter-mixing of poor quality housing.

3.2 Internal structure of the CBD

3.2.1 Within the CBD distinct clusters of functions are often evident giving rise to sub-districts. Herbert (1982) has suggested a model of CBD structure and change over time (Fig. 6). By means of a street survey within the CBD plot and record on a base map the distribution of the six functional categories of land-use shown in Fig. 6.

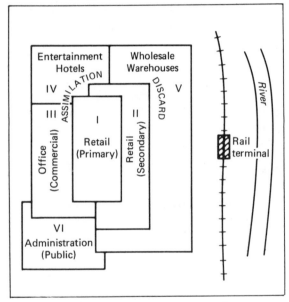

Fig. 6 CBD structure and change over time: a conceptual model (reproduced with permission by D. Herbert)

3.2.2 Variations in land-use are not just found at ground floor level. In properties near the PLVI upper floors will be used for functions unable to afford the higher rents of lower levels. For example, building societies and other finance companies tend to seek central space but are often located above retail outlets. Further away from the centre the number of storeys in buildings will be less, until residential development predominates at all levels. Take a transect extending outwards from the PLVI or centre of the CBD and record the distribution of land-uses making use of upper storeys.

3.3 Changing nature of CBD

3.3.1 A number of studies have noted a process of short-term change by which the CBD was advancing in some directions, the zone of assimilation, and retreating in others, the zone of discard. Generally, the zone of assimilation is towards the higher status residential areas of the city and associated with the growth of new speciality shops, car showrooms, office headquarters and new hotels. On the other hand, the zone of discard is often in the area near to industrial and wholesale activities and by low-status residential districts and is characterised by cheap clothing stores, low-quality restaurants, pawn shops and transport termini (Fig. 6). By field observation plot the presence of these two zones.

3.3.2 It may be useful to combine field observation with an environmental scaling technique. For each shop or office allocate a grade on a scale of 1 to 5 according to its quality. This should be based upon criteria such as, the price and range of goods, type of window display, level of cleanliness and upkeep:

Grades 1 and 2:   high quality
Grade 3:   medium quality
Grades 4 and 5:   low quality

Zones of discard will be associated with low quality scores and zones of assimilation will be related to high quality values.

4. Analysis

4.1 Delimitation of the CBD

4.1.1 Central business index

Draw a map to represent those parts of the central area with a CBHI value of 1.0 or more and CBII value of at least 50 percent. A non-CBD block surrounded by CBD blocks can be included within the CBD as well as public administrative buildings adjacent to the CBD.

Some studies have identified a 'hard core' and 'fringe area' within the CBD using the CB Index. Plot the distribution of those blocks with threshold scores of 4.0 and 80 percent to determine the position of the hard core.

4.1.2 Rate index

On a base map plot the rate index of each property sampled within the central area. These values can then be used as points around which isopleths can be drawn at convenient intervals. Identify the PLVI. Sharp falls in value or clear breaks can be used to distinguish the outer edge of the CBD. A basic problem with this method is deciding the point at which land values cease to be typical of the CBD, inevitably the decision is a subjective one.

4.1.3 Core-frame concept

Draw a map showing those parts of the central area which can be identified as the 'core' and 'frame' according to variations in land-use. The outer boundary of the CBD is usually determined by residential neighbourhoods, heavy industrial concentrations or natural barriers.

4.1.4 Compare the results of any two techniques

In each case measure the size and comment upon the form of the CBD. Consider the relative strengths and weaknesses of each procedure.

4.2 Internal structure of the CBD

4.2.1 Comparison with Herbert's model

Draw a map showing the location of land-uses within the CBD, classified according to the six functional types recognised by Herbert. Functional zones will be evident only in those areas where one particular land-use clearly dominates. Describe the internal structure of the CBD and compare it with Herbert's model.

4.2.2 Vertical distribution of land-use

For the transect draw a diagram to represent the number of storeys per property, distinguishing the vertical variation of land-uses. Is there any variation in upper storey use with distance from the centre of the CBD?

4.3 Changing nature of the CBD

4.3.1 Superimpose a convenient grid over a base map of the central area. For each grid square determine the mean quality score for that area. Consider the distribution of high and low values. Relate these findings to any field observations and attempt to distinguish zones of assimilation and discard.

References

Herbert, D.T and Thomas, C.J (1982) *Urban geography a first approach* (Wiley) London.
Horwood, E.M and Boyce, R.R (1959) *Studies of the Central Business District and urban freeway development* (University of Washington Press) Seattle.
Murphy, R.E and Vance, J.E (1954) 'Delimiting the CBD', *Economic Geography*, 30, 197–223.

8 Urban neighbourhoods

1. Introduction

Generally, urban neighbourhoods are defined in two ways. On the one hand, they are taken to be social areas identified by the neighbourliness and sociability of their residents. On the other, they are taken to be physical units, purpose-built for the provision of housing and services, which in time, it is hoped, will develop social bonds amongst residents.

A basic problem when identifying the former is that it may be very hard for an outsider to distinguish neighbourhood boundaries and form. Yet this sort of information would be very useful to planners in their attempts to create urban units. A possible solution is to attempt to define neighbourhoods through the eyes of their residents.

2. Aims

2.1 To investigate whether residents recognise that they belong to a neighbourhood.

2.2 To examine the influence of socio-personal, functional and environmental factors upon neighbourhood perceptions.

2.2.1 Socio-personal factors

Do such factors as age, sex, marital status, number and age of children, social class, length of residence, potential mobility, the presence or absence of family members and friends in a locality, influence neighbourhood perceptions?

2.2.2 Functional factors

Do the presence or absence of local services such as, schools, libraries, shops, doctors or personal involvement in local affairs through clubs, pubs and politics affect neighbourhood perceptions?

2.2.3 Environmental factors

Does the layout of local roads, housing density, the incidence of physical and man-made barriers such as rivers, railway lines, main roads, areas of open ground, condition neighbourhood perception?

3. Method

3.1 Study area

The project can be carried out in settlements of different size and need not be limited to large urban areas. Ideally, the results from different sorts of areas within a town should be compared. Different criteria can be used to select these areas. For example, location — an inner city area and suburban area; age — new housing development and established residential area; tenure — council housing and owner occupied housing; local services — areas well served and areas poorly served; form — estate and non-estate.

3.2 Neighbourhood perception

Two principal techniques will be used to gain information on neighbourhood perception — a questionnaire survey and a recall mapping exercise.

3.2.1 Questionnaire survey

This provides the means by which socio-personal data about the respondent is acquired and investigates how individuals interact with their local environment, particularly their contact with local services and other organisations. (See Table 11.)

3.2.2 Mapping exercise

Mapping is the technique by which individuals mark the boundaries of their perceived neighbourhood. The procedure involves showing residents a map of the local area upon which their houses are clearly identifiable. Large scale maps at a scale of 1:2,500 or 1:10,000, provide the best framework for this exercise. For each person a clean sheet of tracing paper is placed over the map and the individual is asked to draw around the area considered to be the neighbourhood. Ensure that a cross is placed on the tracing paper corresponding to the home location of the respondent.

An alternative method would be to ask people to name places that form part of their neighbourhood. Such information is translated into a map at the end of the interview.

3.2.3 In order to assess whether individuals are able to perceive neighbourhoods it is important not to lead respondents towards a positive conclusion. Consequently, the term neighbourhood is best avoided early on in the survey, particularly when the interviewer is introducing the topic. Simply point out that a project is being carried out on people's attitudes towards their local area and how they make use of

Table 11

"It is thought that people living in urban areas often make little use of their 'local area' and sometimes know little about those people living around them. I should like to ask you some general questions about the people you know and the facilities you use".

Stress confidential nature of the survey

1. Age of respondent?
   i) < 15   ii) 15–30   iii) 31–45   iv) 46–60   v) > 60
2. i) Male   ii) Female
3. Can you please tell me where you usually go to obtain the following services?

   Location

   i) Butcher
   ii) Chemist
   iii) Post Office
   iv) Doctor
   v) Public House
   vi) Library
   vii) Schools
4. Do you belong to any local organisation of clubs?
   i) Yes   ii) No
5. If Yes, what and where?

   Organisation   Location

   i)
   ii)
   iii)
6. I would like now to consider your close relations for a while, and ask about where they live or last lived?
   i) Grandparents (i.e. where they live or lived — if local specify street address)
   ii) Parents   "   "   "
   iii) Brothers/Sisters
   iv) Children   "   "   "
   v) Grandchildren   "   "   "
7. Having considered your close relatives, this next question changes the emphasis to friends. Could you please tell me where the people you consider as your 'best friends' live? — name no more than five.

   Location

   i)
   ii)
   iii)
   iv)
   v)
8. How long have you lived at this address?
   i) < 1 yr   ii) 1–5 yrs   iii) 6–10 yrs   iv) > 10 yrs
9. Where did you live previously to this address?
10. Can you name your local councillor?
    i) Yes and correct
    ii) No
    iii) Incorrect answer
11. Have you ever written to, or telephoned, a local politician or public official, regarding problems in this area?
    i) Yes
    ii) No
    iii) Do not know
12. Have you ever signed a petition regarding any problem in this area?
    i) Yes
    ii) No
    iii) Do not know
13. Do you feel that you live in a neighbourhood?
    i) Yes
    ii) No
    iii) Do not know
14. If Yes, present with large-scale map, overlay with tracing paper, show and mark position of house.
    Could you please draw the area which you consider to be your neighbourhood on this map?

15. Do you give a name to this neighbourhood?
    If so, what is it?
    i) Yes   ii) No   iii) Do not know
    (If the respondent has said No to Q. 13 ask —
    Do you give a name to this area in which you live?)
16. Can you think of other named areas which join onto your neighbourhood (local area)?
    i) Yes — specify   ii) No   iii) Do not know
17. Present map with neighbourhood marked on it (or if no map, relate to local area). Score the following statements on a scale of 1 to 5.

                            1   2   3   4   5

    i) very important personally — unimportant
    ii) very clearly defined — very poorly defined
    iii) very satisfied with local facilities — very dissatisfied
    iv) very satisfied with neighbourly contact — very dissatisfied
    v) very strong neighbourhood feeling — very weak
18. What is the occupation of the chief wage earner?
19. How many children do you have?
    i) Number < 5 years   ii) Number 5–18 years

places around their home. Also, the mapping exercise should only be introduced if a positive response has been given to the question — 'do you feel that you live in a neighbourhood?'

4. Analysis

Enter the results upon a master sheet; the questions form columns and the answers by each respondent form rows. For some of the questions it is useful to employ a coding system which attempts to assess the importance of the neighbourhood to people's behaviour. Refer to the map drawn by the respondent in response to question 14. Answers to questions 3, 5, 6, 7 and 9 (Table 11) can be coded in the following way: (1) locations inside the perceived neighbourhood; (2) locations outside the perceived neighbourhood; (3) not relevant.

4.1 Neighbourhood perception

4.1.1 Assess the overall percentage of people who recognised that they belonged to a neighbourhood by replying positively to question 13. Provide similar estimates for each area. Do they vary? Do the same for answers to question 14. What do these results suggest about the relevance of the term neighbourhood? How satisfactory is the mapping procedure for assessing neighbourhood size, form and sense of belonging?

4.1.2 Measure the area of each respondent's perceived neighbourhood. This can be assessed by superimposing the tracing paper upon a sheet of millimetre graph paper and counting the squares within the neighbourhood boundaries. This information is then converted to an areal measurement by noting the scale of the original base map upon which the tracing was made. For example, a scale of 1:10,000 would mean that a 1 mm square on the graph paper represents 1 sq metre on the ground. Consider the distribution of neighbourhood sizes, by expressing the mean, median and modal values and calculating the inter-quartile range. Is there any broad agreement amongst the respondents? What does such data suggest about the relevance of the term? For example, if a wide range exists with no apparent grouping of values this would suggest that neighbourhoods have personal relevance rather than group or areal significance.

4.2 Environmental factors

4.2.1 For each area superimpose all the neighbourhood maps outlined on the tracing paper upon the large-scale base map. Ensure that each perceived neighbourhood is centred over the respondent's house. Examine whether people's neighbourhood boundaries are coincident with each other and whether physical properties of the environment influence their delimitation. For example, the presence of a main road forming the boundary to a housing estate may represent a clear edge to many people's neighbourhood maps (Fig. 7).

4.2.2 Consider whether building density influences neighbourhood size. For each respondent count the number of buildings bounded by their perceived neighbourhood, and examine for similarity.

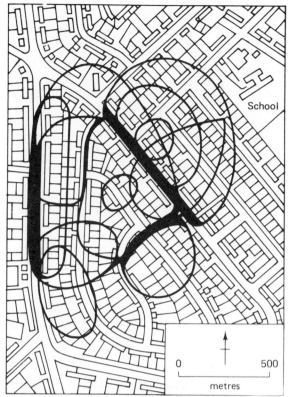

Fig. 7 Superimposed perceived neighbourhood boundaries

4.3 Socio-personal and functional factors

4.3.1 For each area examine the extent to which socio-personal and functional characteristics are found within people's drawn neighbourhoods. For example, what percentage of maps include within them local services, clubs, family members and friends? Are some of these characteristics more important than others?

4.3.2 Do people who draw neighbourhood maps differ to those who do not in terms of these characteristics?

4.3.3 In order to determine whether selected socio-personal characteristics influence neighbourhood perception, the questionnaire and map data need to be sorted according to the chosen variables. For example, the data could be grouped into male and female responses, in order to examine whether sex has a bearing upon neighbourhood perceptions. Other ideas of categorisations are suggested in the 'aims' of this project.

For every male calculate the area of his perceived neighbourhood. Do the same for females. Categorise the results into convenient classes representative of different sizes for example, 0 to 499 m^2, 500 to 999 m^2, 1000 to 1499 m^2, etc. The maps drawn by men and women are then assigned to a class, providing a frequency distribution. The data are now in a form suitable for comparison and chi-squared analysis, enabling the level of association between male and female responses to be assessed. A statistically significant chi-squared value will indicate differences between male and female responses. Suggest reasons why such differences, or, maybe, similarities exist.

4.3.4 Examine the relationships between selected socio-personal characteristics and people's attitudes towards their neighbourhood and local area (Question 17).

4.4 Planning implications

What do your findings suggest about the value of the neighbourhood as a planning goal? For example, are some features, whether environmental, socio-personal or functional, more important to neighbourhood perception than others? How could you use this information to help future urban planning?

References

Jones, E and Eyles, J (1977) *An Introduction to social geography*. (Oxford University Press) Oxford.

Keller, S (1968) *The urban neighbourhood: a sociological perspective*. (Random House) New York.

9 Urban retailing: a behavioural investigation

1. Introduction

Both central place theory and traditional gravity models offer predictions about consumer behaviour. Generally it is presumed that shoppers will be influenced by such factors as distance, size of place and range of goods, always trying to gain commodities with the minimum of effort. Consumers make rational decisions about where to shop, showing a perfect awareness of the availability of different goods.

Increasingly, it is suggested that consumers rarely have complete knowledge of local retailing facilities. Instead, where they shop is the subject of personal decision-making. In order to gain a better understanding of shopping behaviour there is a need to focus upon people's awareness of local shopping centres and to examine those factors which consumers take into account when making a shopping trip.

2. Aims

2.1 To examine patterns of shopping behaviour within an urban area.
2.2 To consider people's perception and awareness of local shopping opportunities.
2.3 To determine those factors considered important by people when deciding where to shop.

3. Methods

3.1 Study area

Within urban areas different types of shopping environment are commonly found. Arranged in increasing order of importance by size, range and kinds of good sold these include neighbourhood, community and regional shopping centres (Fig. 8). If people behaved in a rational way it can be expected that they would travel only locally to obtain daily or low order goods usually to their neighbourhood centre. For irregular or higher order goods they would move slightly further afield to the nearest community or regional outlets.

In setting up this project ensure that the selected urban area contains a range of different sized shopping centres. These should be plotted on a map in order to reveal their distribution. It may be worthwhile contacting the local Town Planning Department at this stage as many of these already hold maps of this kind.

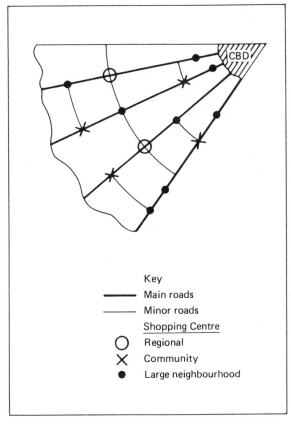

Fig. 8 Intra-urban shopping environments

3.2 Data collection

Recent studies suggest that there is a difference between people's awareness of shopping centres and those which they use. This has led to the recognition of two 'spatial fields':

i) The spatial information field is a zone which includes all those shopping centres about which a consumer has knowledge.
ii) The spatial usage field is a zone which includes all the shopping centres used by a consumer in the course of shopping activities. This zone forms part of the spatial information field.

The main parts of the information and usage fields are shown in Fig. 9.

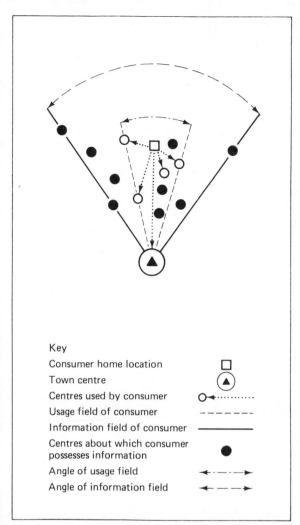

Fig. 9 Consumer information and usage fields

3.2.1 The identification of these two 'spatial fields' provides valuable information upon consumer behaviour and perception of urban retailing. In order to gain such information interviews should be conducted:
1. Ask respondents to name all the shopping centres *within* the town/city with which they are familiar (information field). Make a list of these and if any do not correspond with those plotted on the map, ask the interviewee to locate them for you.
2. Ask respondents to name all the shopping centres *within* the town/city which they use (usage field). Follow the same procedure as above. The questionnaire also needs to explore the reasons why people use certain shopping areas at the expense of others. Attention should focus upon those centres which are most frequently used by the respondent.
3. Ask respondents to name the shopping centre which they most frequently use.
4. Find out what reasons best describe why they visit this centre by showing respondents a list of pre-selected reasons. Yes or No responses should be noted for each of these.
    - i) within walking distance
    - ii) on bus route
    - iii) easy car parking
    - iv) low bus fares/car costs
    - v) quick journey time
    - vi) variety of shops
    - vii) cheap price of goods
    - viii) specialisation of shops
    - ix) good quality shops
    - x) good service
    - xi) near place of work
    - xii) near relatives
    - xiii) see people I know
    - xiv) habit
    - xv) combine with other trips
    - xvi) other reasons
5. Collect information on the following socio-personal characteristics:
    a) age: (i) <25 (ii) 25–39 (iii) 40–60 (iv) >60
    b) car use for shopping: (i) never (ii) infrequently (iii) regularly (iv) always
    c) length of residence: (i) <2 years (ii) 2–5 (iii) >5
    d) occupation of chief wage earner in household.

3.3 Sampling procedure

A variety of sampling procedures may be used. An interesting strategy would be to carry out interviews within three shopping areas, each representative of the different categories shown in Fig. 8. Alternatively, carry out a systematic house to house survey within two areas known to be of contrasting social composition or at different distances from local shopping facilities.

4. Analysis

4.1 For each consumer:

4.1.1 Determine the total number of shopping centres, excluding corner shops, in the information and usage fields.

4.1.2 Plot the angular extent of the information and usage fields. These are obtained, as in Fig. 9, by drawing two lines radiating from the town centre to include all shopping centres within the respective areas.

4.1.3 Calculate the mean distance from home location to shopping centres, for both information and usage fields. By grouping these data together according to different socio-personal characteristics or by shopping centre, some understanding can be gained about people's perception of urban retailing opportunities. For example, the influence of age, length of residence, social class and car usage upon consumer perception and behaviour can be investigated, as well as whether those people visiting regional, community and neighbourhood shopping centres have different perceptions of retail environments. For whatever characteristic under examination discuss the following:

4.1.4 The mean distance from home location to shopping centre for both spatial fields. Do consumers largely travel short distances on shopping trips (distance–decay effect)?; do consumers perceive or prefer to use those centres which are located towards the city centre (downstream directional bias)?

4.1.5 The shape and mean angle of the information and usage fields. Do the angles of these areas vary according to any characteristic? What do these results suggest about people's awareness of shopping opportunities?

4.2 Examine the principal reasons influencing people's decision of where to shop.

4.3 Do the reasons vary according to the characteristics of the consumers or different category of shopping centre? For example, does social class exert an influence upon a person's decision of where to shop?

For parts 4.1, 4.2 and 4.3, consider the implications of these findings with respect to the predictions of central place theory and traditional gravity models, about shopping behaviour. Does this approach provide a better understanding of consumer behaviour?

References

Davies, R (1976) *Marketing Geography* (Retail and Planning Association) Cambridge.

Dawson, J (ed) (1980) *Retail Geography* (Croom Helm) London.

10 Intra-urban industrial change

1. Introduction

In recent decades an important trend has been the decentralisation of industry from the inner-city to suburban locations. Such change has come about through:
i the movement of individual establishments away from the central area;
ii the ability of the outer-urban ring to attract new industries giving rise to a faster growth rate in these areas compared to more central sites. In particular, industrial estates developed on green-field sites have played a major role in influencing the location of manufacturing industry. Associated with these patterns of change are sets of push and pull factors encouraging industries to move, stay or relocate within urban areas.

The principal 'components of change' between two points in time, t_1 and t_2, include:
Entry – an establishment not present in the study area at t_1 but present at t_2. This includes establishments moving into the area from elsewhere and local 'births'.
Exits – an establishment present in the study area at t_1 but not at t_2. This includes establishments moving away from the area to elsewhere and local 'deaths'.
Stayers – an establishment in the same location at t_1 and t_2.
Intra-urban mover – an establishment in a different location in the study area at t_2 from that at t_1.

The ability to distinguish these components is important because each may be associated with different factors promoting stability or change.

2. Aims

2.1   To examine the changing distribution and structure of manufacturing industry within an urban area over time.
2.2   To investigate the principal components of locational change.
2.3   To identify some of the causes of locational change.
2.4   To compare the major characteristics of inner city and suburban industrial structures.

3. Method

3.1   Study area

In a large place select areas where contrast is likely, for example, between an inner city location and a green-field industrial estate. In a small town it may be possible to carry out a full inventory of local manufacturing, but it is still worthwhile distinguishing between an inner and outer zone.

3.2   Time period

The period over which change is to be investigated is largely decided by the availability of local secondary data sources. Ten year periods provide some arbitrary limit, especially as manufacturing seems to be in a state of rapid change. If industrial development grants have been specifically directed at the study area, look at the local area at times previous and subsequent to this legislation.

3.3   Secondary data sources

Only the most accessible data sources are listed below.
3.3.1   Kelly's Directories: This is one of the most useful street directories: information is provided on the name, address and nature of the manufacturing or retail function of an establishment. From 1845 to 1946 Kelly's Directories were published for most counties of the United Kingdom and for most of the largest towns and cities. From 1946 to 1976 Kelly's published annual directories for large towns and cities only. Since 1976 Kelly's has ceased publication except for their Post Office London Directory. Most reference libraries keep copies of these directories, although the earliest publications are most likely to be stored in the local archives office.
3.3.2   Telephone directories: Both Yellow and White Page Post Office Directories are useful as data sources, especially for checking dates of openings and closures and identifying local movement. Additional information on ownership and status can then be added by reference to 'Who Owns Whom', published since 1958 by Roskill and Co. (Reports) Ltd.; 'Kompass', published annually since 1962 by Kompass Publishers Ltd.; Key British Enterprises and Kelly's Manufacturers and Merchants Directory.

Supplementary sources which may be available at a local level include the Industrial Market Location Directory, Dun and Bradstreet UK Market Facts File, as well as data collected by local authorities. Many such bodies carry out their own censuses of industrial location which provide considerable information on local establishments.

3.3.3 Direct surveys: Street surveys, whereby information is collected by 'walking the streets', provides the most satisfactory means by which to compile an inventory of local manufacturing businesses. In order to assist analysis plants should be listed according to the Standard Industrial Classification (S.I.C.) developed for the Census of Production (1980) (Table 12).

3.4 Questionnaire survey

In order to attempt explanation of the changing patterns of manufacturing interviews with management are invaluable. Students are reminded not to be too ambitious in terms of sample size. A starting point is to classify establishments according to whether they are entries, movers, or stayers — for exits it may prove impossible to trace local 'deaths' but transfers and branch closures may be contacted through their parent company. A stratified random sample reflecting the relative proportion of each category in each area can then be conducted.

Interview programme. Much background information upon each establishment, including details of ownership, can be gained from secondary sources. Attention should focus upon the characteristics of the establishment and the factors influencing locational decision-making.

Table 12 Standard Industrial Classification (1980)

21	Extraction and preparation of metalliferous ores
22	Metal manufacturing
23	Extraction of mineral ores
24	Manufacture of non-metallic products
25	Chemical industry
26	Production of man-made fibres
31	Manufacture of metal goods
33	Manufacture of office machinery and data processing equipment
34	Electrical and electronic engineering
35	Manufacture of motor vehicles and parts thereof
36	Manufacture of other transport equipment
37	Instrument engineering
41/42	Food, drink and tobacco manufacturing industries
43	Textile industry
44	Manufacture of leather and leather goods
45	Footwear and clothing industries
46	Timber and wooden furniture industries

Objective data on the establishment. Length of time operated at this location; previous location, if any; other locations of firm; product types; employment levels and composition (e.g. employment ratios of male to female, skilled to unskilled; part-time to full-time); market (location of main business clients).

Factors encouraging present location. What caused you to open an establishment in this location? This question may be left 'open' or structured around a series of 'closed' alternatives. For example, availability of space; low rents; inadequate existing premises or site; permit an expansion of output; assisted area grants or loans; proximity of other firms; good accessibility; good parking or unloading facilities; immediate occupance of factory buildings; desire to be in more attractive surroundings; good for image/improved status; closer to markets; security; good internal road system; nearness to supplies; others, please specify. Respondents can be asked to tick those factors which actively influenced their decision-making or even to number these in order of importance.

Factors promoting change. Give the main circumstances which led to the move away from your last location? For example, congestion; lack of room to expand; poor road system; difficulty of accessibility; town planning difficulties; lack of security; desire to be in far more attractive surroundings; too far from suppliers/markets; others, please specify. The main consideration when selecting these factors is to distinguish between those 'push' and 'pull' forces which encouraged the movement of establishments and contributed to industrial change. Additionally, for those establishments classified as 'stayers', try to establish what factors encouraged this stability and explore whether their production or employment characteristics changed over the specified period.

4. Analysis

The accompanying notes presume that comparison is to be made of two areas of contrasting location and industrial character.

4.1 For each point in time, t_1 and t_2, draw a dot map to represent the location of each manufacturing establishment. Between t_1 and t_2 how many establishments have been gained or lost in each area? Calculate the percentage of change for each area and distinguish between absolute and relative growth or decline. Is there any evidence for decentralisation of industry?

4.2 It is useful to sort the net changes into components of change. For each area, between t_1 and t_2 examine the number and proportion of entries, stayers and intra-urban movers. These data can be represented in diagramatic form (Fig. 10) or mapped.

4.3 With reference to the S.I.C., what sorts of establishments are opening, closing, moving or staying within each area? What are the places or origin of intra-urban movers? Is there any relationship between patterns of change and the availability of local industrial grants?

Rank in order of number of establishments the

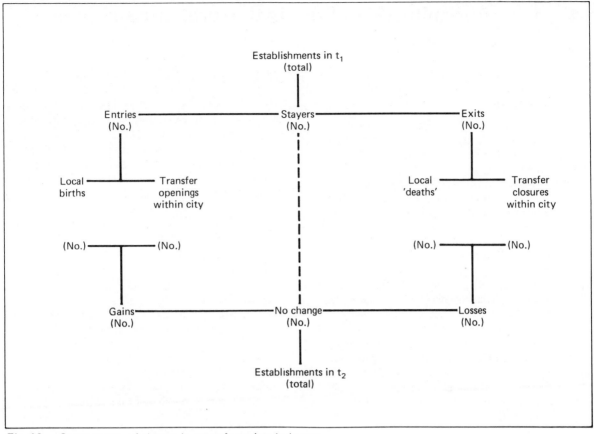

Fig. 10 Components of change in manufacturing industry

manufacturing activities associated with each area for t_1 and t_2. Has the overall structure of manufacturing activities changed over time?

4.4 'Who Owns Whom', 'Kompass' and 'Key British Industries' distinguish between those establishments which are head offices or branches, and independents or subsidiaries. Has the pattern of ownership changed within areas between t_1 and t_2? Is there any evidence for concentration into large or small companies?

4.5 Questionnaire analyses. This should attempt to summarise some of the factors influencing locational change, as well as highlighting the characteristics of those establishments associated with different components of change.

4.5.1 Classify establishments according to the main components of change (entries, exists, stayers and movers). Examine the main circumstances which encouraged openings, stability or movement into or away from each area. Contrast those factors associated with inner city and suburban locations. Try to derive a classification which summarises the principal factors promoting change or stability; for example, push, pull or inertial reasons.

4.5.2 For each general component of change, compare and contrast the objective data derived upon each establishment. This can be carried out for the sample as a whole and for each area. For example, are the employment characteristics or markets of those establishments which opened in an area different to those which survived? What proportion of each category have been in receipt of industrial grants? In this way it should be possible to establish whether entries, exits, stayers and movers differ in terms of these characteristics and this should assist explanation of change.

References

Healey, M.J (ed) (1983) *Urban and regional industrial research: the changing UK data base* (Geo Books) Norwich.

11 Agricultural land-use in the rural–urban fringe

1. Introduction

An interesting area in which to carry out an investigation of agricultural land-use is that around the edge of urban centres, sometimes termed the rural–urban fringe.

Three schools of thought have emerged to explain agricultural land-use patterns around urban areas:
(i) Traditional (direct) approach: this approach is closely associated with the ideas of Von Thunen. He suggested that the intensity of land-use decreased in concentric zones with increased distance from the urban place.
(ii) Modern (reverse) approach: according to Sinclair a reverse of Von Thunen's pattern is typical, so that in the outer fringe agricultural land-use is likely to be more intense than within its inner zone.
(iii) Behavioural approach: recent evidence suggests that urban pressure can produce both negative and positive responses from agriculture within the rural–urban fringe, such that land-use can be of varying intensity dependent upon farmers' perceptions and their decision-making behaviour.

2. Aims

2.1 To consider the intensity of agricultural land-use within the rural–urban fringe.
2.2 To examine the applicability of traditional (direct) and modern (reverse) approaches towards explaining agricultural land-use variations in the rural–urban fringe.
2.3 To explore those factors which may influence farmers' decision-making within the rural–urban fringe, especially their perception of the effect of urban pressure upon agriculture.

3. Method

3.1 Study area

The framework for this investigation is the parish, as considerable agricultural data exists for this administrative unit. The Ordnance Survey publishes parish maps at a scale of 1:100,000.

The first step is to attempt to define the limits of the rural–urban fringe around an urban centre. Inevitably, its boundaries will be difficult to locate. Usually, the inner fringe extends up to 1.5 km away from the urban edge where land is in a state of transition from rural to urban. The outer fringe covers a broader area of 1.5 to 10 km in width and is an area where rural land dominates, but some urban elements are still apparent. Outline a transect extending from the inner to outer edge. The breadth of the transect depends upon the shape, form and number of local parishes. A target of no more than 15 parishes forms a realistic sample.

3.2 Data collection

3.2.1 Parish summaries

Annual parish summaries provide an important data source for agricultural statistics. These contain information on the number of farms found locally, their size and speciality, the amount of farm labour, the quantity of livestock and the total hectares devoted to different land-use categories such as crops, grains and horticulture. These are collated by the Ministry of Agriculture, Food and Fisheries (M.A.F.F.), Guildford and stored for public use at the Public Records Office, Kew. Copies of the parish summaries can be purchased for a small cost. Local Agricultural Development and Advisory Service (A.D.A.S.) offices also hold parish summaries for their respective county divisions.

3.2.2 Intensity index

Agricultural intensity can be measured by considering labour inputs within each parish. From the parish summaries record the amount of land devoted to arable farming and the quantity of livestock. The values for each category are converted to 'standard man day equivalents' by reference to the table of Standard Labour Requirements (M.A.F.F.) (Table 13). A 'standard man day' (S.M.D.) approximates to an 8 hour working day and thus, is an indicator of labour intensity. For example, the amount of labour required for growing one hectare of wheat is equivalent to 2.5 working days. If 80 hectares of wheat are grown in the parish this is equivalent to 200 S.M.D.'s (i.e. 2.5 × 80). On the other hand, if 120 cows are kept in the parish, the S.M.D. rate is 7.0 per head amounting to 840 S.M.D.'s (i.e. 7.0 × 120). By adding together the results for each variety of crop and type of livestock a parish total is obtained, which when divided by the size of parish expresses local agricultural intensity:

$$\text{Agricultural intensity} = \frac{\text{Total parish 'S.M.D. equivalents'}}{\text{Parish size}}$$

Table 13 Extracts from Standard Labour Requirements 1976 (M.A.F.F.)

	Standard man days per hectare
Crops and Grass	
Wheat, Barley, Oats, Rye	2.5
Potatoes, early	25.0
Potatoes, main crop	30.0
Turnips	15.0
Cabbage, Savoys for stock feeding	12.5
Bare Fallow	1.0
Horticultural Crops	
Carrots, for market	135
Carrots, main crop	20
Parsnips	17
Onions, for salad	200
Onions, for harvesting	35
Broad Beans, for market	35
Broad Beans, for processing	5
Green Peas, for market	75
Green Peas, for processing	5
Horticultural Crops	
Roses	300
Dessert apples	60
Strawberries, open grown	150
Livestock	
Cows and heifers in milk, for producing	
milk	7.0
beef	2.5
Bulls for service >2 years	6.0
<2 years	2.5
Sheep, >1 year	0.5
Pigs	3.5
Poultry, growing pullets	0.04
Ducks	0.1
Turkey	0.08

3.2.3 Farm survey

Land-use variation within the rural—urban fringe may bear little resemblance to that described by the reverse or direct approaches. The farm survey focuses upon individual farmers within the study area. The purpose of the questionnaire is to gain an understanding of agricultural decision-making by considering farmers' perceptions of the effects of urban pressures. (See Table 14.)

In order to undertake this survey a small sample of farms should be selected. Either, choose a constant proportion of farms from each parish or focus upon those parishes which deviate most from predicted patterns and compare these with others which conform to expectations.

4. Analysis

The parish summaries provide information upon agricultural intensity and farming trends.

4.1 Agricultural intensity

Calculate the total scores for the intensity index. Either:
i) draw a choropleth map of intensity scores for the parishes; or
ii) plot a graph, representing intensity values on the vertical (y) axes, against distance from the urban edge horizontal (x) axes. Compare the results with the patterns predicted by the direct and reverse approaches. Is there any evidence of zonation?

4.2 Farming trends

Three other farming characteristics can be determined from the parish summaries: the number of farms, their size and levels of diversification. Draw maps to represent these data and consider their variation within the rural—urban fringe. Do farm sizes increase with distance from the urban edge? Is specialisation more typical of those farms nearest the urban centre?

4.3 Farming characteristics

For selected information derived from the questionnaire an impression of farming within the rural—urban fringe can be gained. For each parish consider such detail as: (a) farm size variations; (b) incidence of full-time and part-time farming; (c) proportion of crops and livestock; (d) levels of land transference; (e) impact of land transference on farm structure. In this case, distinguish between negative (Q.10.i—iii) and positive (Q.10.iv—vi) responses; (f) farm fragmentation. Maps and diagrams will assist description. For example, divided proportional circles could be used to represent different farm sizes and their proportion of crops and livestock.

In all cases consider whether distance from the urban edge leads to any variation amongst these farming characteristics. Consider the implications of these findings with respect both to the agricultural models and towards providing an understanding of farming behaviour within the rural—urban fringe.

4.4 Farmer perception

What factors seem to influence farmers' perceptions of the threat of urban pressure? Have different sorts of farmers responded in different ways to these perceived pressures? Do perceptions towards urbanisation vary according to such factors as location or farmer status? Have farmers changed their attitudes and perceptions? Are farmers closer to the urban fringe more pessimistic about urban pressure than those further away?

Table 14 Questionnaire for farmers

A. *Farm and farmer characteristics*
1. Age   i) < 30   ii) 30–45   iii) 45–65   iv) > 65
2. What practical agricultural training have you received?
   i) none   ii) family farm/father   iii) farm pupil apprentice   iv) day release   v) agricultural college
3. How long have you been a farmer?
   i) < 5 years   ii) 5–10 years   iii) 11–20 years   iv) > 20 years
4. Which of the following best describes your status on the holding?
   i) owner occupier   iv) leaseholder
   ii) owner and tenant   v) manager
   iii) tenant   vi) other
5. Are you a (i) full-time or (ii) part-time farmer?
6. What crops/livestock are on your holding?
   Please state whether these are increasing (I) decreasing (D) remaining unchanged (U) in the last 5 years. Place enterprises in rank order of profit.

Crop/Livestock	Hectares	I/D/U	Rank
i)			
ii)			
iii)			
iv)			

7. Has any land been transferred from agricultural to non-agricultural land use?
   i) Yes   ii) No   iii) Do not know
8. If Yes, was this land transfer   i) voluntary or   ii) the result of a compulsory purchase order?
9. How many acres were transferred and when?
10. What effect did the loss/sale of the land have on your farm?
    i) made it less profitable   ii) caused you to change crops and/or livestock
    iii) caused you to cease full-time farming in favour of part-time farming   iv) allowed you to spend the money from the sale on capital items such as machinery or buildings   v) allowed you to spend the money on land purchase elsewhere   vi) had little effect   vii) other effect
11. Is your land in (i) one block or   (ii) fragmented?
12. If fragmented has this been due to any kind of urban pressure?
    i) Yes and specify why   ii) No and specify why

B. *Farmer perception*
13. Have your crops/property been subjected to any of the following forms of interference by the general public? If so, how serious was this interference?
    Seriousness on Scale (1–5) (Not serious-extremely serious)

i) Trespassers	Yes	No	Scale _____
ii) Broken fences	Yes	No	Scale _____
iii) Dogs	Yes	No	Scale _____
iv) Litter	Yes	No	Scale _____
v) Vandalism	Yes	No	Scale _____
vi) Other	Yes	No	Scale _____

14. Has this interference had any effect on your farming operations?
    i) Yes   ii) No   iii) Do not know
15. If yes, what?
16. Do you feel that your farm is threatened by urban pressure of any kind?
    i) Yes   ii) No   iii) Do not know
17. If yes, what?
18. What is the likelihood that your farm will lose more than 10 per cent of its area within the next 10 years to urbanisation?
    i) very unlikely   ii) possible   iii) probable   iv) very probable   v) a certainty
19. In the next ten years will agriculture remain economically viable in the parish?
    i) Yes   ii) No   iii) Do not know

References

Ilbery, B.W (1978) 'Agricultural decision-making: a behavioural perspective' *Progress in Human Geography*, 2, 448–66 (Edward Arnold).

Ilbery, B.W (1985) *Agricultural Geography* (O.U.P.)

12 Service hierarchy and spheres of influence

1. Introduction

Central Place Theory is an attempt to explain the numbers, sizes and distribution of service settlements. Two concepts are especially important to its ideas: threshold population and range of goods. Threshold is the minimum demand necessary to ensure a commodity or service is offered at a centre; range is the maximum distance consumers are prepared to travel to obtain a commodity or service. Services are not uniform in their thresholds and ranges and thus will serve and draw upon varying hinterlands (complementary regions).

Additionally it is assumed that economic forces influence behaviour: the entrepreneur is keen to derive maximum profit from the sale of goods and the consumer will try to obtain these services with the minimum of cost and effort.

The combination of these economic forces has considerable spatial implications: (i) Within any area there will be many low order, fewer middle order and even less high order facilities. (ii) A structured arrangement of central places will develop in order to offer these services. (iii) Each central place will serve a specific hinterland or complementary region. The hinterlands of lower order centres will be arranged within those of higher order. (iv) These central places will be evenly spaced over the landscape, with the lowest order places closest together and the highest furthest apart.

2. Aims

2.1 To consider whether settlements show a structured arrangement when measured and ranked by the size and variety of their service goods.

2.2 To consider whether a structured arrangement of complementary regions is apparent.

2.3 To examine whether settlements display a uniform, clustered or random spatial pattern.

3. Method

3.1 Study area

Ensure that the area chosen for investigation displays a range of settlements with different size and service characteristics. In order to decide which settlements to include in the survey a possible strategy would be to carry out a stratified random sample within the defined area. To do this, categorise all settlements according to their population size (e.g. 0–499; 500–999 etc.). The 1981 census will provide this information. For each category use random number tables to choose a specified percentage of settlements, for example, 10% of each group. In this way a sample is derived which is representative of settlements of varying size.

3.2 Data collection

3.2.1 Measures of the importance of central places

Centrality expresses the importance of a central place. High centrality refers to a place with a wide range of high and low order goods; conversely, low centrality relates to a settlement with few low order facilities. Commonly, two approaches have been used to determine centrality:

(i) For the first technique, decide upon a set of functions which characterise different orders of goods. A coding scheme can be used which awards points for the incidence of these types of function within a centre. High order services are given more points than those of lower orders e.g. low order functions 1 point, middle order functions 2 points, high order functions 3 points. Each centre is awarded a total number of points according to its complement of facilities. For example, for *each* of the following within a settlement award, 1 point: convenience food store, public house, church, post office, sports facility, petrol station, guest house etc. 2 points: chain food store, primary school, doctor, bank, hotel etc. 3 points: supermarket, department store, secondary school, hospital, theatre. Most of this information can only be gained by walking around each centre, although the Yellow Pages telephone directory should help to locate some of the services.
(ii) For the second technique within each settlement record every service outlet by its dominant line of business. These data provide the basis of an index of centrality which expresses the importance of a service centre.

3.2.2 Measures of complementary regions

Two major approaches have been employed to distinguish the scale and extent of complementary regions, these can be termed indirect and direct.

Indirect method: This attempts to determine complementary regions by plotting the trade areas of selected facilities offered at or distributed from

settlements. Choose a range of facilities and determine their catchment areas from each appropriate settlement within the study area. For example:
(a) delivery area of selected retail firms (b) areas served by insurance offices and banks (c) school catchment areas (d) distribution and reporting areas of local newspapers (e) areas served by local buses.

Direct method: This involves interviewing consumers from different parts of the study area to determine which centres they visit to obtain retail and service facilities. Choose a range of facilities representative of different orders of good and ask consumers where they would obtain these items. For example,
Shopping – convenience food, furniture, clothes, gramophone records, jewellery, footwear, electrical goods, hardware.
Professional – accountant, dentist, schooling, solicitor, optician, insurance, bank, chemist.
Entertainments – cinema, dance, library.

4. Analysis

4.1 Measures of centrality

4.1.1 Coding scores

(i) Produce a list of settlements arranged according to their total scores. Are there any obvious groups of settlements with similar values? Are there more low order than middle and high order settlements?
(ii) Plot a graph of coding score of each settlement (x axis) against population size (y axis) of each settlement. Does this graph reveal any groups of settlements? Draw a best-fit line by eye through the distribution of points. Which places deviate most from this line?

To what extent do these results confirm the ideas of Central Place Theory?

4.1.2 Service inventory

(i) Plot a graph of number of activities in each centre (x axis) against population size of each centre (y axis). Are there any obvious groups of settlements? If groups are evident compare the range and types of function which they offer.
(ii) Calculate the centrality status of each centre by using the location quotient:

$$C = \frac{t}{T} 100$$

- t = one outlet of function t
- T = total number of outlets of function t in the study area
- C = centrality value

This method presumes that the centrality (focality) of a function varies with the total number of establishments of that function within the study area. For example, if there is only one outlet for a function within the region it can be presumed that this will draw upon a wide hinterland; whereas if there are lots of outlets for a function these will have small hinterlands. 'C' expresses the extent of focality: the higher its value the greater is the focality. There are two steps to its calculation.

Step one: for example, if there are 200 bakers in an area the location quotient is:

$$C = \frac{1}{200} \times \frac{100}{1} = 0.5$$

Step two: if in settlement A there are 23 bakers the centrality value for that function is

$$0.5 \times 23 = 11.5$$

By totalling all the centrality values for every function within a settlement an index is obtained which expresses the degree of focality of that place. The higher the index the more important the place.

List each centre by index value. Consider the arrangement of centres. Centres which have similar values are classed members of the same group. Is there a clear, structured pattern?

4.2 Complementary regions

4.2.1 Indirect method

(i) Map the catchment areas of each facility and examine their pattern and distribution. Around each settlement, particularly the larger ones, can a broad hinterland be recognised? Can regions be distinguished by the coincidence of different catchment boundaries? Is there any evidence for nesting, that is, do the hinterlands of the smaller centres tuck inside the wider hinterlands of the larger centres?

4.2.2 Direct measurement

(i) The results of the questionnaire can be used to distinguish complementary regions. The hinterlands of a settlement may be divided into two zones:
(a) an intensive area within which consumers patronise the central settlement for at least 50 per cent of all goods.
(b) an extensive area which shows the maximum area to which the influence of the central settlement extends for a service.
Do complementary regions overlap? Is there a clear structured arrangement of complementary regions?
(ii) The questionnaires also provide information on which centres consumers use for particular goods.

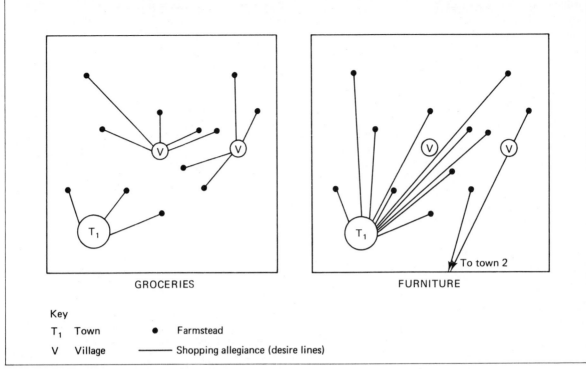

Fig. 11   Journey to shop movements

For selected goods draw 'desire lines' extending from the place of residence of a consumer to the place at which the good or function is obtained (Fig. 11). Calculate the mean distance travelled by consumers to reach different goods. What do these results suggest about the hinterlands of different goods and settlements?

4.3 Spatial characteristics

Nearest-neighbour analysis provides a means to examine the spatial characteristics of the settlement pattern.

(i) For each order of central places measure the distance from each settlement to its nearest neighbour of the same grade. Calculate the mean nearest-neighbour distance for each order. Are central places of the lowest order closer together than those of the highest order?

(ii) Theoretically, there should be a uniform distribution of central places. Within the real world this is unlikely but the settlement pattern should still display a tendency towards uniformity. The nearest-neighbour measure (RN) provides a description of the settlement pattern in terms of whether it is clustered, regular or random. Any calculated value for RN will fall on a continuous scale from 0 (clustered) to 1.0 (random) to 2.149 (regular).

(a) For each settlement measure the distance to its nearest-neighbour regardless of order

(b) Add all the distances and divide this total by the number of measurements to obtain the observed value $\bar{D}o$

(c) Calculate the mean value for a random distribution:

$$\bar{D}r = 0.5 \sqrt{\frac{A}{N}}$$

N = number of settlements

A = area of study area

(d) Calculate R_N: $R_N = \dfrac{\bar{D}o}{\bar{D}r}$

Relate this nearest-neighbour value to the theoretical scale and test for significance.

References

Blunden, J, Haggett, P, Hamnett, C, and Sarre, P (1978) *Fundamentals of human geography* (Harper and Row) London.

Bradford, M.G and Kent, W.A (1977) *Human geography* (Oxford University Press) Oxford.

13 Routine meteorological observations and microclimatic studies

1. Introduction

Weather data have been recorded for more than 200 years in the United Kingdom but, for many areas of the world, observations at ground level stations have been made for considerably less than 100 years. Meteorological measurements are needed because weather plays such an important role in our economic, social and sporting activities and we therefore need much better methods of weather forecasting. In recent years, more and more data are collected electronically, especially from fixed orbit satellites.

Ground radar stations also play an important role in forecasting, by tracking rainstorms and providing data for early floodwarning, for example. Despite these developments local meteorological observations are still important for both verifying remotely sensed information and for providing data on localised conditions.

2. Aims

i) To describe methods for measuring meteorological information at a fixed station.
ii) To describe methods for measuring meteorological data in remote locations.
iii) To investigate the urban environment as a suitable location for microclimatological studies.

3. The meteorological station

The meteorological station should be located in an area of open ground as far away from buildings, trees and fences as possible. As a rough guide, the station should be located at a distance of at least twice the height of any obstruction or building. Having selected the site, an area of 10 m × 6 m is required for the location of all recording instruments as shown in Fig. 12. All readings should be taken at daily intervals, usually at 9 o'clock in the morning and data should also be collected at week-ends and during school vacations whenever possible. Readings may also be taken at much more frequent intervals for specific experiments.

Fig. 12 Layout for a meteorological station

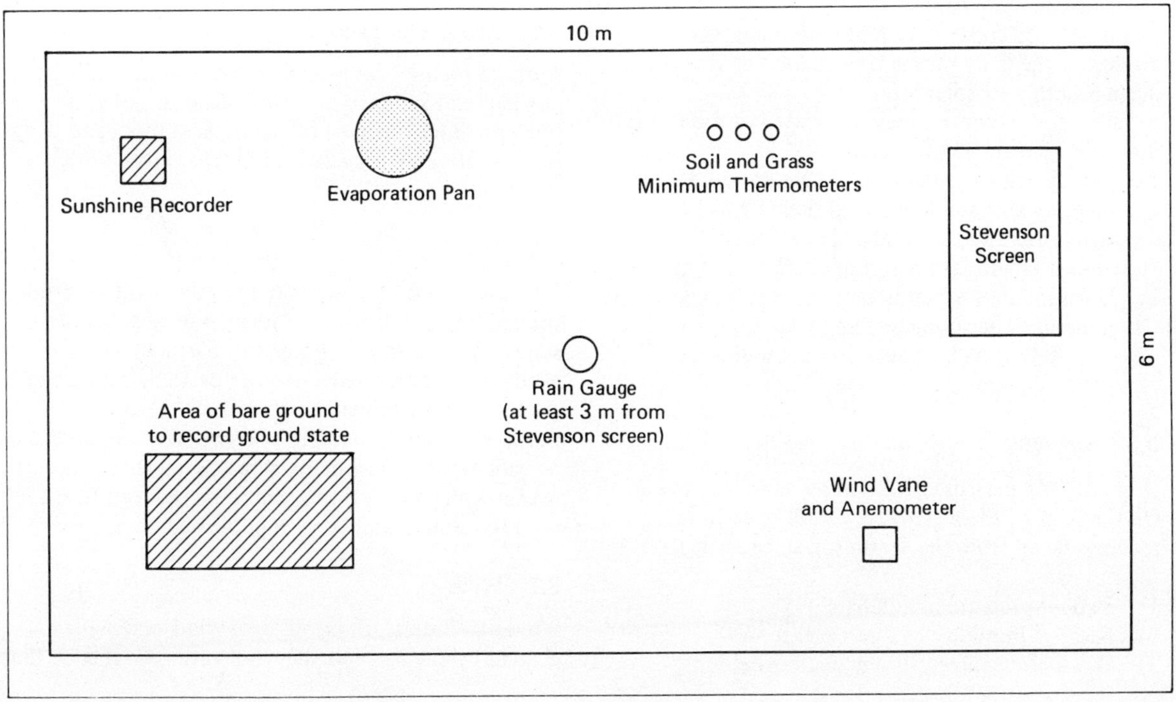

3.1 The Stevenson screen

Many recording instruments are sensitive to direct sunlight and should be kept in a shaded position within a Stevenson screen. The screen is mounted on a frame approximately 1½ m above ground level, and arranged so that the door opens in a north-easterly facing direction.

3.2 Temperature measurements

The Stevenson screen will normally contain a minimum of four thermometers for measuring shade temperatures under different conditions. The first two thermometers are designed to measure the maximum and minimum temperature. The former is mercury filled and pushes a small indicator upwards as the mercury expands from the bulb. The 'minimum' thermometer is alcohol filled because the indicator is set within the fluid which therefore needs to be transparent.

The second pair of thermometers are the 'wet' and 'dry' bulb thermometers which are usually mounted on a small stand. These are both mercury filled, but one is covered by a cotton bag which is connected via a wick to a water filled container and must be kept moist at all times. The two values are used for humidity calculations as described in section 3.3. The following results should be noted.

1) Mean Daily Temperature — average of minimum and maximum value for the day.
2) Daily Range — difference between the maximum and minimum value.

For each calendar month of recordings, the average of the daily means is calculated to give the mean monthly temperature.

Other thermometers may be located outside of the Stevenson screen. These include:—
1) Soil—Earth thermometers. These are simple mercury thermometers with their stems bent at right angles and are embedded in the soil to depths ranging from around 2 cm to 20 cm.
2) Grass minimum thermometers. This is an alcohol minimum thermometer supported on Y shaped pegs so that its bulb is in contact with close cropped grass.

3.3 Humidity

Humidity is a measure of the water vapour in the air. On the basis of the wet and dry bulb temperature readings taken from the Stevenson screen calculate the following:
1) Vapour pressure in millibars
2) Relative humidity as a percentage
3) Dew-point temperature in degrees celsius

These values may be calculated using a humidity slide rule, or, more accurately, from the Hygrometric Tables prepared by the Meteorological Office (H.M.S.O. 1964) which are based on the Regnault equation. Different sets of tables are used depending on whether Stevenson screen readings or whirling hygrometer readings are being used (see section 4).

3.4 Precipitation

Rainfall measurement, at least in principle, is very simple. It is recorded as the depth to which rain or its liquid equivalent has accumulated through an orifice of precisely known area. In the UK, the rainfall is usually measured in a gauge of 12.7 cm (5 in) diameter and 30.5 cm (12 in) high. Where commercial equipment is not available, polythene funnels of similar dimension, mounted at 30.5 cm above ground level and leading into a plastic bottle, may prove adequate.

A raingauge is of little use for the measurement of snow depth because the gauge will fill rapidly with even light falls. Two methods may be used to estimate snowfalls. First, an open container of 20 cm in diameter and 20–40 cm high may be used to collect falling snow. This is removed to the classroom and allowed to melt before the volume of the liquid is measured. Alternatively, snow depths over the meteorological station may be measured with a wooden rule at approximately 10 locations, and the average snow depth recorded (the ratio of snow depth to water equivalent is approximately 10:1).

3.5 Evaporation and transpiration

A direct method for assessing this component is given in Chapter 15, but in meteorological stations, an open pan of 122 cm (4 ft) diameter is used, and daily levels (corrected for rainfall input) are read with a micrometer gauge.

3.6 Pressure

Pressure is normally measured in millibars to 0.1 mb using a mercury barometer located in the Stevenson screen. Several types of mercury barometer are available, some of which require periodic calibration. Alternative instruments include the Aneroid Barometer and Barograph. The two instruments must be calibrated against mercury barometers for absolute readings and all pressures should be reduced to sea level for comparability with published data.

3.7 Wind

Wind direction is measured by a wind vane which is aligned by the force of the wind acting on it such that it is pointing into the wind. Wind speed is measured by an anemometer. In meteorological stations, the voltage generated by rotation of the cups is recorded as a continuous measure of wind speed but hand held anemometers are also available and can be used to good effect for daily observations by timing the number of revolutions over a minute and calculating wind speed from a calibration equation (supplied by manufacturers).

3.8 Sunshine

The number of hours of sunshine received at a ground station must be recorded continuously; usually on a light or heat sensitive paper. The standard instrument in the UK is the Campbell-Stokes recorder, but simpler, and less expensive models use slit holes and light sensitive paper, such as the Jordan sunshine recorder.

3.9 Additional routine observations

The meteorological measurements described in the previous 8 sections should be supplemented by a written description of other weather conditions. Notes should be made on the following at the time of observation:
1) Cloud cover. The proportion of the sky covered by cloud is reported on the standard Octa scale as given in Table 15.
2) Cloud type and height. Clouds should be classified into their major types and the approximate height of the cloud base recorded.

3) Visibility. Visibility readings depend on the establishment of objects at different distances from the meteorological station. Up to 10 objects, ranging in distance from 55 yards to over 31 miles (where possible) should be noted and used to classify visibility as shown in Table 15.
4) State of bare ground. This is a simple description of the condition of the bare ground plot on the meteorological station (e.g. wet, dry, snow covered).
5) Note should also be made of weather conditions at the time of observations (e.g. raining).

4 Mobile meteorological observations

For microclimatic studies, it is important that the instruments are mobile, and that conditions similar to those at the meteorological station are met in order to provide comparative data with the base station. Not all meteorological variables will change sufficiently in small areas to justify measurement, such as atmospheric pressure, since any large change in this parameter would suggest a change in air mass. However, the following variables can be recorded easily for microclimate studies.
1) Shade temperature
   Shade temperature of the air can be recorded on a thermometer in a small portable Stevenson screen or box which is painted white in order to reflect direct radiation. This should be held at the desired recording point for 1 minute to allow the thermometer to respond to local temperature.
2) Wet and dry bulb temperature
   These temperatures, for humidity calculations, can be measured with a whirling hygrometer. This comprises two thermometers mounted in a frame similar to a football supporters rattle. One thermometer is moistened from a water filled reservoir and the second records dry air temperature. The hygrometer is whirled at head height for 1 minute and the temperature readings noted. Calculations for humidity follow those of section 3.3, but the Hygrometric Tables calibrated for Aspirated Psychrometer readings should be used in place of the tables for Stevenson screen readings.
3) Wind
   Both wind speed and direction can be measured in microclimate studies. Observations can be made with portable hand held weather vanes and anemometers similar to those described in section 3.7.

Table 15 Recording of cloud cover and visibility data

Cloud cover									
Cover	Nil	$\frac{1}{8}$	$\frac{2}{8}$	$\frac{3}{8}$	$\frac{4}{8}$	$\frac{5}{8}$	$\frac{6}{8}$	$\frac{7}{8}$	$\frac{8}{8}$ Sky	8 Obscured*
Code	0	1	2	3	4	5	6	7	8	9

* By dust or fog etc.

Visibility										
Distance	55 yd	220 yd	550 yd	1100 yd	$1\frac{1}{4}$ mil	$2\frac{1}{2}$ mil	$6\frac{1}{4}$ mil	$12\frac{1}{2}$ mil	31 mil	over 31 mil
Code	0	1	2	3	4	5	6	7	8	9

Code	Description	Code	Description
0	Dense fog	5	Poor visibility
1	Thick fog	6	Moderate visibility
2	Fog	7	Good visibility
3	Moderate fog	8	Very good visibility
4	Mist or haze	9	Excellent visibility

5. Application in urban microclimate studies

The construction of urban areas leads to the destruction of one microclimate and the creation of a new one. Urban areas have an effect on atmospheric composition, on the heat budget and on surface roughness and hence airflow patterns and characteristics.

Atmospheric composition changes are difficult to measure, because they include aerosols gases and particulate dust emissions. However, both modification of the heat budget and changes in airflow characteristics can be measured successfully in the urban environment at different scales.

At the macroscale a transect through an urban environment, measuring the parameters described in section 4, can be used to examine humidity, temperature and airflow characteristics in three dimensions. Transects should be completed within 1–2 hours at most because of possible air mass changes and diurnal changes in humidity and temperature characteristics. Where a number of students are involved, several routes through the urban environment may be selected to provide a spatial picture of microclimate characteristics. Where possible, an attempt should be made to gain access to tall buildings in different parts of the urban area, and measure microclimate changes with height. Several surveys should be undertaken either under different air mass conditions or at different times of the day or night. All data should be presented graphically or in map form.

At the microscale, changes in climate may be measured around individual buildings. In this case, access to different levels of the building for measurement is essential. Attempts should be made to record differences in temperature, humidity and wind speed which should then be mapped on detailed plans of the individual buildings.

References

Barry, R.G and R.J Chorley (1982) *Atmosphere weather and climate* 4th ed. (Methuen).

McIntosh, D.H and A.S Thom (1973) *Essentials of meteorology*. (Wykeham).

Meteorological Office (1964) Hygrometric tables.

Part II (2nd ed) *Stevenson Screen readings Degree Celsius* HMSO met. 0.265B.

Part III (2nd ed) *Aspirated Psychrometer readings Degrees Celsius* HMSO Met. 0.265C.

14 Forest interception studies

1. Introduction

Under most conditions, precipitation falling on the land surface does not reach the ground directly, but is caught or intercepted by the leaves, branches, trunks and stems of growing vegetation. Not all of the water falling above a forest canopy will reach the forest floor, but a proportion will be caught in small leaf and bark depressions. The volume retained is the *interception capacity* of the vegetation. In coniferous forests, this capacity may vary very little throughout the year but in deciduous forests, the interception capacity will be greatest during the period of maximum leaf development in the summer.

In addition to affecting the volume of water which reaches the ground surface, a vegetation canopy will also alter the distribution of precipitation of the forest floor. Where water is intercepted by trees, some will drip from branches and leaf tips by a process known as throughfall and can be intercepted again at a lower level by herbs, shrubs and grasses by a process known as secondary interception. Some water may run along leaves and branches and reaches the ground surface by flowing down tree trunks by a process known as stemflow. Both stemflow and throughfall may occur at different levels within a vegetation canopy, especially in woodlands where there are usually at least two levels of vegetation development (tree and ground vegetation). Any water which remains in the interception store after rainfall stops may be lost by direct evaporation from the foliage. These processes affecting the redistribution of water beneath a vegetation canopy are shown schematically in Fig. 13.

2. Aims

The aims of this exercise are:
i) To illustrate a simple method for collecting throughfall in a forest.
ii) To examine a variety of sampling designs which may be used to test different hypotheses regarding the throughfall process.

3. Designing throughfall collectors

Throughfall is considerably more difficult to measure with any degree of certainty than rainfall because the volume of water dripping through a vegetation canopy varies over very small distances. It is therefore possible to overestimate throughfall volumes by positioning the collector immediately beneath a point of leaf drip or underestimate throughfall by missing drip points entirely. For this reason, it is difficult to use the standard Meteorological Office 12.7 cm (5″)

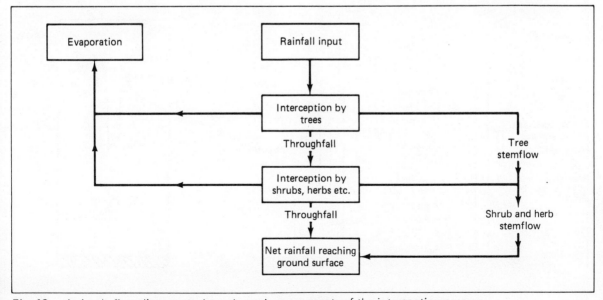

Fig. 13   A simple flow diagram to show the main components of the interception process

diameter gauge because it samples a very small area. Where they are used, sampling will have to be undertaken at a very high density in order to obtain an accurate estimate of average throughfall volume. A better solution to this problem is to increase the area of the throughfall sampler, and this can be done by using 1 m lengths of plastic guttering, mounted at an angle of 22°, on wooden posts hammered into soil. The steep angle is required to minimise the problem of outsplash from the gauge. At the lower end of the gutter, the throughfall is collected in a 5 litre polythene bottle via a plastic funnel plugged with glass wool to prevent leaves and twigs from entering the collector (Fig. 14a).

Since the volume of throughfall collected is, like rainfall, measured over a horizontal projection of the sampling area, the length of gutter over which throughfall is collected is:

$\quad$ Cos α x gutter length

where α = angle of gutter

At an angle of 22° and with a length of 1 m, the length is Cos 22° $\times$ 1.0 m = .9272 m. The surface area over which throughfall is collected is gutter width $\times$ length: for a 10.16 cm width gutter

= 10.16 cm $\times$ 92.72 cm

= 942.04 cm^2

Fig. 14 Simple design for
 a) A gutter throughfall collector
 b) A funnel throughfall collector

Where throughfall gutters are too expensive or too difficult to make, large diameter (approximately 23 cm:9 in) plastic funnels of known surface area can also be used (Fig. 14b). These have less than half the sampling area of the gutter collectors, and should therefore be installed in the field at approximately twice the sampling density in order to produce comparable results. Their main advantage is that they are cheap and much easier to install for short term experiments.

All of the samplers described above may be made at low cost, but will provide data sufficiently accurate for the testing of many different hypotheses and relationships.

4. Sampling designs for hypothesis testing

A variety of simple projects may be designed around the funnel or gutter throughfall collectors. The two experiments outlined below are illustrative and can be modified or combined with other projects to very good effect. In all cases, precipitation in an area of open ground should be measured in close proximity to the sampling site.

4.1 Throughfall variability around a single tree

At the small scale, the volume of throughfall reaching the ground surface will decrease as samples are collected closer to the tree trunk and volume increases in a roughly concentric pattern with

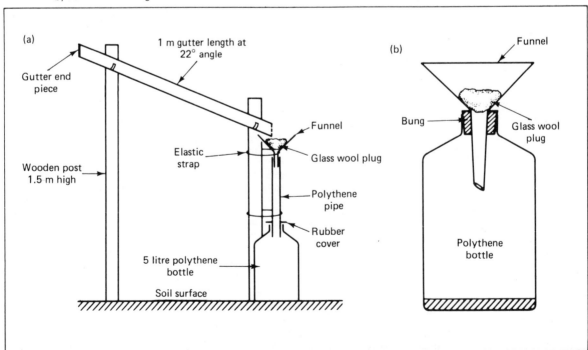

increasing distance from the trunk.

Select two to three large isolated trees, either of different species and similar age or of the same species and similar age or of the same species but different ages, and arrange funnel throughfall collectors in the sampling scheme shown in Fig. 15a. The distance between each concentric ring should be approximately 0.5 m, and should extend to the edge of the tree canopy. It is important that single isolated trees should be selected for this experiment because interference will occur when tree crowns overlap. The volume of water collected in each funnel and bottle should be measured at daily or weekly intervals or after each individual rainfall event.

Calculate the mean throughfall volume for each distance from the four funnel collectors and plot average volume against distance from trunk. Plot similar graphs for the different tree species on the same graph and compare the results. The average throughfall volume for each tree can also be calculated by summing the individual volumes and dividing by the number of collectors. This data may be used to estimate the interception capacity of the tree crown as described in section 4.2

Fig. 15. Sampling strategies to establish:
    a) microscale variations in throughfall catch around a single tree
    b) throughfall volumes of a small area using i) Funnel collectors ii) Gutter collectors

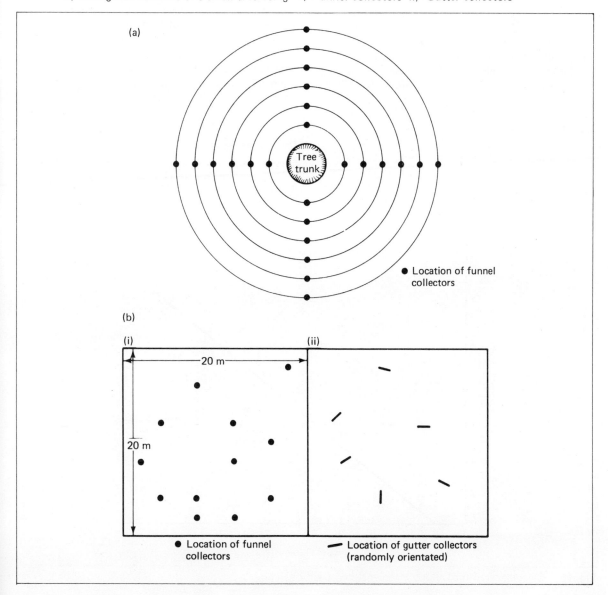

4.2 The calculation of throughfall variability and interception capacity

Calculation of throughfall volumes for a given area of woodland can be undertaken by selecting a fairly homogenous area and laying out a sampling grid of approximately 20 m by 20 m as shown in Fig. 15b) i) Where funnel collectors are used, approximately 12 of these should be located using random coordinates in a pattern similar to that shown in Fig. 15b) ii). Where gutter collectors are used a smaller number are needed (approximately 6) but in addition to locating the sampling points randomly the gutters should be orientated in random directions in relation to the coordinates of the sampling grids shown in Fig. 15b) i). The orientation can be selected from random number tables and fixed in the field by compass bearing. Again, samples may be collected at weekly intervals or after each major rainfall event. The following statistics should also be calculated.

i) Mean throughfall volume for the area
ii) Coefficient of variation (%) of throughfall volume:—

$$CV (\%) = \frac{\text{Standard Deviation}}{\text{Mean}} \times \frac{100}{1}$$

Where throughfall samples have been collected in areas with different tree species or trees of different age, the Mann-Whitney U test may be used to establish whether the throughfall volumes are significantly different for the two areas.

Calculation of the interception capacity of the tree canopy must be undertaken by plotting the relationship between precipitation input to the area and average throughfall volume as shown in Fig. 16. Draw a best fit line through the relationship by eye, and find the point on the rainfall axis at which the best fit line shows a throughfall of 0.0 mm (labelled A on Fig. 16). This value is approximately the interception capacity of the tree crown. Similar relationships may be drawn for different times of the year to establish whether the interception capacity changes during the growing season for example.

5. Additional comments

The experiments outlined above may be conducted independently or, where a large number of students are working on the problem, a group project combining both aspects may be successfully designed. The projects may be extended by including microclimatological studies, such as air temperature, soil temperature and relative humidity measurements, for comparison with meteorological data collected from a permenent field site.

References

Gregory, K. J and Walling, D.E (1976) *Drainage Basin Form and Process*. (Edward Arnold).
Smith, D.I and Stopp, P (1978) *The River Basin*. (Cambridge).

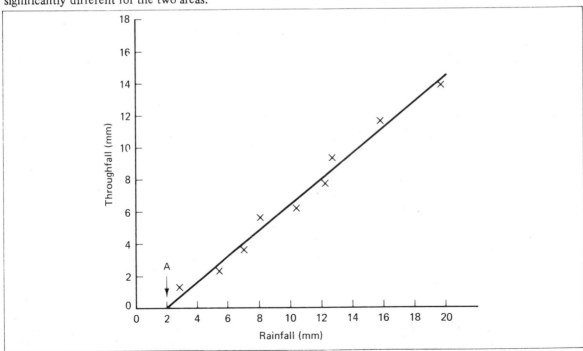

Fig. 16 Calculation of the canopy interception capacity

15 Measurement of the water balance in soil profiles

1. Introduction

The water balance of a soil profile describes the input, output and storage changes which take place in a unit of soil of known surface area and volume. It can be thought of in terms of an accountants book balance because the total sum of all the inputs should equal all the outputs plus or minus the change in the amount of water held within the account.

Because the soil profile acts as a filter for incoming precipitation, the change in soil moisture content will be controlled by:—
1) The previous soil moisture content
2) The amount of water falling onto the soil surface
3) The amount of water lost by evaporation and transpiration
4) The amount of water lost by drainage to the groundwater zone.

2. Aims

The aim of this chapter is to demonstrate a simple technique for measuring the water balance of a soil profile and to illustrate the application of such measurements in meteorological and hydrological studies.

3. Constructing a lysimeter

Water balance measurements for soil profiles are made with percolation gauges or lysimeters. Many different types exist and the basic design given in Fig. 17 shows a simple weighing lysimeter. It is a cylinder of approximately 25 cm in diameter and 40–50 cm in depth made of metal or strong plastic. The top is open to allow water entry from precipitation and the bottom end drilled with holes to allow free passage of water. Tubular steel for construction is expensive and, unless stainless, will rust. However, large tin-plated cans of approximately similar dimensions are used for catering purposes by many school kitchens and would be adequate for this purpose. Two holes should be drilled on either side of the lysimeter, at the open end, into which hooks can be fitted for removing the lysimeter from the soil profile. When fully saturated, a container of the dimensions shown in Fig. 17 could weigh over 50 kg! A second cylinder of slightly larger diameter but of much greater length should also be obtained and should form a waterproof seal to prevent the inflow of water from the surrounding soil profile when buried. This container should be deep enough to hold a polythene bottle and collecting funnel supported immediately beneath the base of the lysimeter as shown in Fig. 17. Under ideal circumstances, the lysimeter should be filled with the undisturbed soil obtained from the hole excavated for its emplacement. This is particularly difficult and therefore every effort should be made to excavate the soil with minimal disturbance and repack it into the cylinder in the correct sequence. The turf or vegetation should be cut with a sharp trowel and removed as a unit before the remaining soil is removed. This turf should be carefully fitted into the lysimeter so that it is flush with the top of the lysimeter and surrounding soil surface.

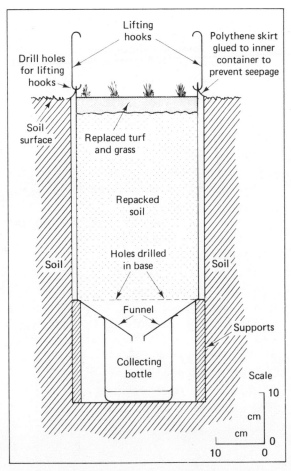

Fig. 17 A simple design for a lysimeter

In addition to the lysimeter, the following equipment is required.
1. A spring balance or set of scales capable of weighing up to 60 kg ± 5 g (Greengrocers scales may be used to good effect).
2. A raingauge to measure precipitation inputs to the soil profile (see chapter 13).

3.1 Experimental design

Installation of a single lysimeter allows calculation of a water balance for the profile. However, a second cylinder can be constructed to act as a control for specific experiments. In this case, for example, vegetation could be destroyed by application of a granulated or powdered herbicide which would prevent growth. This simple test would enable the difference between evaporation from bare soil and vegetated soil to be calculated. If a herbicide is added in liquid form, remember to note the volume of liquid applied and to account for this in the water balance calculation given in section 3.2. Once installed after initital weighing, the lysimeter should be removed and weighed periodically and the volume of drainage water and precipitation should be measured. This can be done on a weekly or monthly basis or after each major rainfall event. On the basis of these three measurements, the water balance can be calculated as shown below.

3.2 Example water balance calculation

1. Dimensions of lysimeter
   Surface area = πr^2 Diameter = 25 cm
   Radius = 12.5 cm
   Surface area = 490.81 cm^3
   Volume = $\pi r^2 h$ Depth = 40 cm
   Volume = 19632.4 cm^3
2. Initial weight of lysimeter plus soil = 45.502 kg
3. Recorded Data after one week
   Rainfall (mm) – 5.0
   Drainage water (cm^3) – 8.0
   Weight of lysimeter plus soil (kg) – 45.593
4. Calculate moisture content change in lysimeter
   Weight change = recorded weight – initial weight
       (kg)   (kg)   (kg)
   .091 = 45.593   – 45.502
   ∴ weight change = + 91 g
   Assume 1 cm^3 water weighs 1 g
   ∴ the increase in soil water volume is 91 cm^3

5. Calculate % moisture content increase by volume
   $= \dfrac{\text{Volume added to soil water}}{\text{Lysimeter volume}} \times \dfrac{100}{1}$
   $= \dfrac{91}{1963204} \times \dfrac{100}{1}$
   $= + 0.46\%$

6. Convert measured volumes to depth equivalents. Since the rainfall depths are in mm and drainage and moisture contents are in cm^3, they must be converted to the same units to solve the water balance equation. By convention, all units are expressed in terms of rainfall depth, in mm. The surface area of the lysimeter is 490.81 cm^3
   Therefore, Drainage (mm) = $\dfrac{8}{490.81}$ = 0.016 cm = 0.16 mm
   Similarly, the increase in soil moisture,
   (mm) = $\dfrac{91}{490.81}$ = 0.185 cm = 1.85 mm

7. Solve the water balance equation to calculate evaporation and transpiration (ET).
   ET = Precipitation – (Drainage + increase in soil moisture)
   ET = 5 – (0.16 + 1.85)
   ET = 2.99 mm

3.3 Presentation of results

Plot graphs or histograms of the following information at the appropriate time interval on graph paper.
1. Precipitation
2. Changes in soil moisture content in the lysimeter
3. Rates of evaporation and transpiration

Where a control lysimeter is used, 2 and 3 should be plotted for both lysimeters on the same graph for comparison.

Plot the cumulative precipitation and cumulative evapotranspiration for comparison of the water balance over long time periods.

References

Knapp, B.J (1979) *Elements of Geographical Hydrology* (Allen and Unwin).
Smith, D.I and Stopp, P (1978) *The River Basin* (Cambridge).

16 The measurement of river velocity and river discharge

1. Introduction

River discharge is defined as the volume of water flowing through a given point in a river channel over a known period of time. For large rivers, the measurement unit is the cumec (cubic metre per second). The Amazon, for example, has a mean annual discharge of around 180 000 cumecs whereas the mean annual discharge of the Thames is only 60 cumecs. For much smaller rivers, discharges are measured in litres per second ($1 s^{-1}$), where $1000 \, 1 s^{-1}$ is equal to 1 cumec. River discharge may be measured by a number of methods, but the most widely applied requires an estimate of the average velocity (in $m \, s^{-1}$: metres per second) and measurement of the cross-sectional area of the flowing water (in m^2: square metres). Discharge (in cumecs) is the product of velocity and area.

The techniques outlined below are designed to measure velocity and discharge in small rivers with discharges of less than 2–3 cumecs and usually less than 1 cumec.

2. Aims

1. To measure the velocity of a river.
2. To measure river discharge.
3. To explore a small number of field projects which may be undertaken with reference to velocity and discharge measurements.

3. The measurement of river velocity

Two basic methods are normally used to determine river velocity. The first requires the timing of surface floats placed in the stream and allowed to run over a measured distance. The second, and more precise technique, requires use of a current meter (which necessitates the purchase of more expensive equipment).

3.1 Surface floats

The simplest form of float measures the surface velocity of the river. Experimental investigations of river velocities have shown that the average velocity of a vertical section is 0.85 times the surface velocity. Velocity decreases with depth because of the frictional drag between the flowing water and stream bed. Velocity decreases towards the banks of the stream for the same reason. If surface velocity is measured by the use of floats, the average velocity of the stream can be estimated by multiplying surface velocity by 0.85.

The float should be constructed so that it travels completely submerged in the flow. It should be relatively short, usually less than 10 cm, and should not be used where minimum water depths are less than 15–20 cm. Floats may be made using small plastic screw top bottles, and filled with sand or soil so that they just become submerged in the flow. This is important, because high winds across the river surface will affect velocity measurements.

3.1.1 Procedure

1. Select a reach of river between 10 and 50 m in length. This reach should be straight and free from surface vegetation, growing weeds and large boulders interrupting the flow. The reach, under ideal conditions, should have a stable cross-section (i.e. no excessively deep or shallow sections).
2. Drop a float into the centre of the stream, at least 3 m upstream of the upper measuring point, and record the time it takes for the float to pass between the marked beginning and end of the reach.
3. Repeated runs (approximately 5) should be made over each reach. Ignore any measurements where the float touches the river bank or an obstruction in the stream channel.
4. Tabulate the results, and calculate the average surface velocity and corrected river velocity as shown in Table 16.

Table 16 Calculation of river velocity using floats

Measured Distance = 10 m			
Time Taken (secs)	Average time (secs)	Average surface Velocity ($m \, s^{-1}$)	Average river velocity ($m \, s^{-1}$)[1]
63			
61			
64	62.2	0.1608	0.1367
62			
61			
1 Calculated from average surface velocity × 0.85			

Unless experiments are undertaken in rivers greater than c 1–2 m in width, it will not be possible to measure variations in velocity across the river cross-section using this method.

Measurements of average velocity for a given river reach may also be used to calculate river discharge as shown in section 4.1.

3.2 Current metering

A more accurate estimate of river velocity, and hence river discharge, can be obtained by using a current meter. This is a small propeller mounted on a wading rod and pointed upstream parallel to the main flow direction (Fig. 18). The number of revolutions of the propeller is proportional to the velocity of the stream at the measurement point. Each current meter and propeller is laboratory calibrated and reference should be made to the calibration equations supplied by the manufacturer.

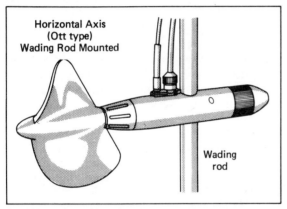

Fig. 18 An Ott type current meter for flow gauging

The velocity is measured at a number of points across the channel cross-section, and at different depths within the profile, by timing the number of revolutions of the propeller over 60 seconds and converting the number of revolutions per second into a measure of velocity. When measurements of velocity are only being used to calculate discharge, divide the width of the stream section into 11 equal distances (to give 10 equally spaced points of measurement across the stream) and measure the velocity at a distance up from the stream bed of 0.4 × water depth as shown in Fig. 19. For example, at point 4 on Fig. 19a), the water depth is 30 cm. The average velocity is measured at 0.4 × 30 cm or 12 cm above the stream bed, as shown on a field booking sheet in Table 17. Average velocity for the river is found by summing the individual velocities for each measuring point and dividing by the number of points. Discharge for the section is calculated by the method given in section 4.2.

Table 17 Recorded cross-section and velocity data from field survey

Width of water surface = 110.0 cm Distance between measuring points = $\frac{\text{width}}{11.0}$ = 10cm			
Distance from left bank (cm)[1]	Water depth (cm)	0.4 × depth[2] (cm)	Velocity (ms^{-1})
10	20	8.0	.061
20	26	10.4	.079
30	28	11.2	.083
40	30	12.0	.108
50	31	12.4	.157
60	31	12.4	.164
70	30	12.0	.158
80	29	11.6	.136
90	20	8.0	.068
100	10	4.0	.047

1 looking upstream
2 for measuring velocity by current meter

Where an experiment has been designed to examine variations in velocity across a channel cross-section, ten equally spaced vertical profiles may again be used, but on this occasion velocity should be measured at three depths corresponding to 0.2, 0.4 and 0.8 times the depth of the stream above the river bed at each section. For example, with a 30 cm water depth, velocity should be measured at 6 cm, 12 cm and 24 cm above the stream bed. The velocity measurements can be plotted as shown in Fig. 19b) on a scaled cross-section of the stream channel, and lines of equal velocity (isovels) constructed by eye at suitable intervals for the channel cross-section (Fig. 19c)). Detailed measurements of velocity distributions in closely spaced sections around meander bends or in pools and riffles (deep and shallow) stream sections will enable the identification of the change in the point of maximum velocity through a small reach of river.

4. The measurement of river discharge

When discharge measurements are made at a river section using the velocity area technique, the cross-sectional area of the stream is also needed because discharge is the product of area and velocity.

4.1 Discharge calculations using surface floats

For the stream cross-section, measure the width of the water surface (in m) and divide by 11.0 to give ten equally spaced points across the stream. Stretch a tape measure across the stream and at each point measure the water depths and note them in a field notebook as shown in Table 17 for the cross-section plotted in Fig. 19a). These data can be used to draw a scaled cross-section of the river on graph paper. By using the same vertical and horizontal scales on the

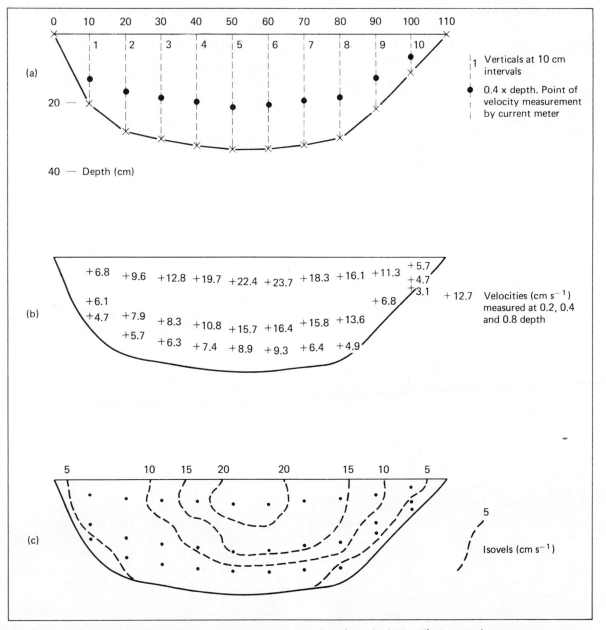

Fig. 19 a) Location of verticals and velocity measuring points for calculating discharge using the velocity-area technique b) Velocities (cm s^{-1}) measured in the same section at 0.2, 0.4 and 0.8 × water depth for constructing isovels. c) Isovel diagram (Isovels join points of equal velocity) from data in b.

graph paper, count the number of squares enclosed by the cross-section and calculate the true area in square metres.

4.1.1 Example

With a cross-section drawn from the tabulated data in the first two columns of Table 17, the total area of flowing water is 0.2565 m^2. With an average velocity of 0.1367 m s^{-1} taken from Table 17, the discharge is

0.1367 × 0.2565 = 0.03506 m^3 s^{-1}

Since a cumec is 1000 litres per second, the discharge may be reported in litres per second (1 s^{-1}).

0.03506 × 1000 = 35.06 l s^{-1}

Units of litres per second should be used where discharges are less than 1 cumec.

4.2 Discharge measurements using the current meter

The calculation of discharge using the current meter initially follows the procedure of section 4.1 by drawing an accurate scaled cross-section, as shown for the first four readings of Table 17, in Fig. 20. Each point of velocity measurement represents the mid point, in terms of area, of a section of flowing water. The area A_1 on Fig. 20 is associated with the volocity V_1, and so on. From the scaled profile, calculate the area of each section; multiply each area by the respective velocity and calculate discharge in $1\ s^{-1}$. From Table 17 and Fig. 20;

Area (m²)		Velocity (m s⁻¹)		Discharge (1 s⁻¹)	
A_1	1.022	V_1	.061	Q_1	1.34
A_2	.025	V_2	.079	Q_2	1.98
A_3	.028	V_3	.083	Q_3	2.32
"	"	"	"	"	"
"	"	"	"	"	"
"	"	"	"	"	"
"	"	"	"	"	"
A_{10}		V_{10}		Q_{10}	

Total discharge for the stream is found by adding the ten discharge values together from each stream segment.

5. Sample projects

Measurement of river discharge and river velocity provides a basis for the investigation of many properties of rivers. The projects briefly outlined below are included to illustrate only the more general principles and could be undertaken in isolation or in combination.

5.1 Velocity changes in a river system

Many early ideas of fluvial processes suggest that rivers flowing in upland 'youthful' areas were characterised by high velocities. This was generally believed because the gradient of upland rivers is much higher than lowland rivers. Although gradient in part controls stream velocity, other factors such as the roughness of the channel and the proportion of the flow volume in contact with the bed and banks of the stream reduce flow velocities because of increasing friction.

These hypotheses may be tested by measuring average river velocity for a small river basin from its source downstream by taking velocity measurements every km for a distance of c 10–20 km. It is important that similar sites should be chosen for these measurements. Many channels show alternating deep (pool) and shallow (riffle) sections, and it is suggested that all measurements should be taken at the mid point of riffle sections for comparative purposes. The velocity (and discharge data if also measured) can be plotted against distance downstream or against drainage basin area contributing to the point of measurement, to see whether velocity or discharge increases or remains constant in a downstream direction.

5.2 Velocity variations through meander bends

This project can only be undertaken with reference to accurate current metering techniques as described in section 3.2. For a single meander bend system, select 5–6 cross-sections and accurately record velocity at three depths and at ten points across the river. Plot scaled cross profiles and map the isovels as shown in Fig. 19c) for each cross-section. Plot the position of the maximum flow velocity on a sketch plan of the meander bend to establish whether the line of maximum velocity is in the mid point of the cross-section or whether it changes position through the meander bend. This project could be extended by an analysis of the particle size distribution of sediments in the channel bed following the techniques discussed in Chapter 20.

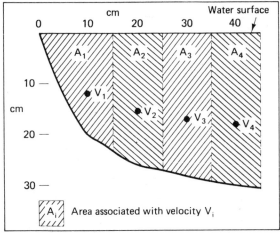

Fig. 20 Calculation of segment area (Ai) and associated segment velocity (Vi) for determining river discharge by the velocity area method

References

Gregory, K.J and Walling, D.E (1976) *Drainage Basin Form and Process* (Edward Arnold).
Smith, D.I and Stopp, P (1978) *The River Basin* (C.U.P.)

17 Morphological, geomorphological and slope mapping

1. Introduction

Investigation of the physical landscape often requires the production of a large-scale map which shows land form (the morphological map), physical features (the geomorphological map) and/or slopes (the slope map). These investigations are important for describing specific features or for simplifying an apparently complicated set of features produced by several processes. Sources of information on the physical environment include the published Ordnance Survey map, as well as specialist geological, hydrogeological, soil and land-use maps. However, the scale and contour interval chosen for all of these maps means that much information of interest to the geomorphologist is omitted, even on the 1:2500 Ordnance Survey maps. Although a study of large-scale aerial photographs may partly solve these problems, it is difficult and expensive to derive information on slope angle, for example, even though some of the more obvious physical features may be identified.

Field mapping can be undertaken with a minimum of equipment, using published Ordnance Survey maps at an appropriate scale as a base map. Information of interest to the geomorphologist is plotted on the base map, which contains enough information to enable location and orientation of specific features in the field. The technique does not require precise surveying techniques, and the end result is an annotated sketch map.

2. Aims

To illustrate a technique which will be used:-
1) To produce a simple morphological/slope map and/or geomorphological map with a defined field study area (c. 5–8 km²).
2) To produce similar maps which concentrate on erosional or depositional processes within a clearly defined environment (e.g. coastal, fluvial or glacial).

3. Methods

3.1 Choice of base-map scale

One of the most important considerations in conducting field mapping is the selection of a base map at a suitable scale. Careful thought should be given to the size and area of individual features to be mapped, and the size of the region. For example a drumlin 50 m long would only be 1 mm in length on a 1:50,000 OS map. A guide to the selection of the most appropriate map scale is given in Fig. 21 which shows the change in length and area of a 100 m × 100 m feature between the 1:50,000 and 1:2500 OS maps. Usually the most useful base maps for field mapping are at the 1:10,000 (6″) or 1:1200 (25″) scale.

Fig. 21 The effect of length and size of feature on the choice of map scale

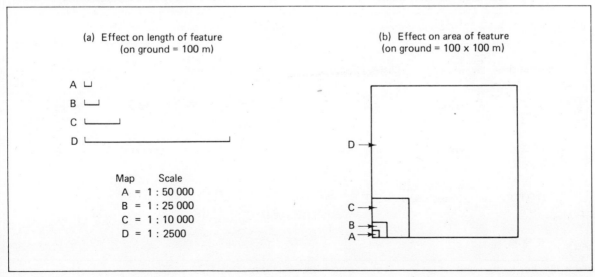

3.2 Morphological and slope mapping

Morphological mapping is based on the assumption that the landscape can be divided into a number of easily identifiable categories for which specific symbols should be used to aid interpretation. These include convex, concave and rectilinear slope sections (See Fig. 22). Changes in slope can be either sharp (i.e. a break of slope) or gradual (a change in slope) and should be identified on the map with the appropriate symbol (Table 18). In some cases, a slope facet is identified where there is a convex slope above changing to a concave one below, and a specific symbol is used to portray this change (Table 18).

These divisions and symbols give rise to 6 major morphological units: —

Morphological Unit	Description
1) Slope facet	Sloping plane
2) Flat	Level surface
3) Curved Unit	Area with constant profile curvature
4) Concave Unit	Curved unit with negative profile curvature
5) Convex Unit	Curved unit with positive profile curvature
6) Cliff	Slopes over 40° in bare rock

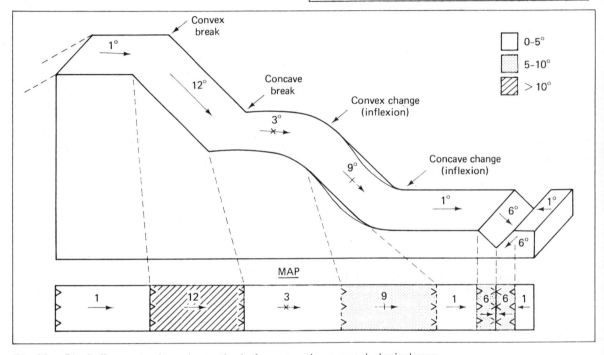

Fig. 22 Block diagram to show the method of constructing a morphological map

Table 18 Key for the identification and use of morphological mapping symbols

In addition to the identification of breaks and changes in slope, arrows should be added to the map to indicate angle of true slope:—

———5°——→ Rectilinear slope facet
———X——→ Convex slope facet
———1——→ Concave slope facet

Slope angle should be measured with an inclinometer, abney level or slope pantometer.

The morphological map is often difficult to interpret, and a simple shading should be employed to clarify the identified slope units (See Fig. 22).

Glacial and Periglacial Features (light blue)	Fluvial Features (dark blue)	Coastal Features (green)
Cirque	Stream	Surf zone
Glacial trough	River channel	Drift direction
Glacial drainage channel	Dry channel	Rock platform
Drumlin	Waterfall	Beach ridges
Roches Moutonnées	Rapids	Abandoned cliff
Terminal moraine	Sand bar	Offshore bar
Lateral moraine	Oxbow lake	Spit
Medial moraine	Point bar	Beach ridges
Esker	Alluvial fan	Lagoon
Kame deposit	Delta	**Slope Features (brown)**
Outwash	Marsh	Landslide
Patterned ground	Lake	Rock fall
Stone stripes	Dry valley	Debris fan
		Mudflow
		Solifluction lobe
		Soil creep

Table 19 Some typical symbols for use with geomorphological maps

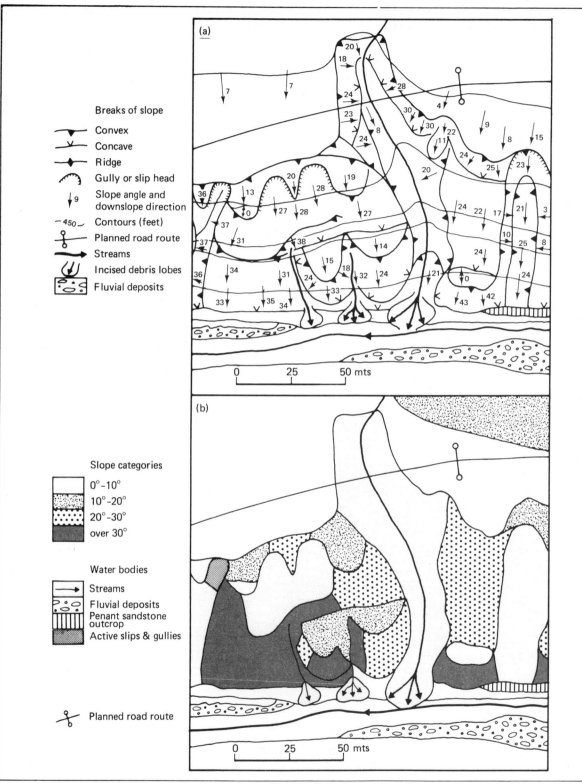

Fig. 23 Morphological a) and Geomorphological b) map of mass movement features in a valley section in South Wales

3.3 The geomorphological map

The geomorphological map is usually constructed in association with the morphological map and is used to show specific features using a system of coloured shading and standardised symbols. These include glacial and periglacial features (light blue), fluvial features (dark blue), karst features (orange), mass movement features (brown), coastal features (green), aeolian features (yellow), volcanic features (red), structural features (purple) and man-made features (black). Many published standard guides are available which provide detailed genetic or classificatory mapping symbols for different environments (e.g. Cooke and Doornkamp, 1974; Gardiner and Dackombe, 1983). A simplified set of symbols for use in fluvial, glacial, coastal and mass movement studies is given in Table 19 and an example of a morphological and geomorphological map is given in Fig. 23.

4. Field survey

Field survey should be conducted with considerable care, paying special attention to features and slope units large enough to be plotted, and using information on the base map as a guide to location and orientation. Base maps should be mounted on a solid backing, such as a clipboard; notes and sketches of the area should be made in a soft pencil and a large polythene bag should be used to protect the map in bad weather. It is often useful to conduct a reconnaissance survey of the area prior to the construction of the map. This allows initial identification of features of interest and may be used to modify the sampling strategy for map construction. A systematic survey of a given area is recommended for the inexperienced mapper. This may be achieved by traversing large areas of the map on pre-determined compass bearings, a procedure which ensures uniform coverage. Slope angles should be measured across all morphological units. If a slope pantometer is not used, it is convenient to measure angle with an inclinometer or abney level over a 5 m distance.

5. Analysis and interpretation

Production of the morphological and/or geomorphological map is not an end in itself. Like most field and laboratory techniques, it is used as an aid to understanding the physical environment. Given the diverse range of problems to which the technique may be applied, it is neither possible nor desirable to lay down strict guidelines relating to interpretation. However, the description of the area in question could follow the aims set out in section 2.

Emphasis could be placed not only on the mapped landforms and features but also on the processes responsible for their development. Simple analyses of the frequency distribution of mapped slope angles could be made, and the map could be supplemented by field notes and detailed sketches of features of specific interest.

References

Cooke, R.U and Doornkamp, J.C (1974)
Geomorphology in environmental management (Oxford University Press) Oxford.

Gardiner, V and Dackombe, R (1983)
Geomorphological field manual (Allen and Unwin) London.

18 Analysis of hillslope erosional processes

1. Introduction

Hillslope erosion takes place through a variety of agencies, some of which are so slow as to be imperceptible to the naked eye (such as soil creep) and some take place at speeds approaching the velocity of rivers (such as mud-flows). Two basic methods are available to examine hillslope erosional processes. First, the resulting landform features, such as mass movements, rotational slips and rockfalls can be mapped by the methods described in chapter 17. Secondly, some attempt may be made to measure the processes directly.

2. Aims

To examine methods for measuring rainsplash and overland flow erosion by means of carefully controlled laboratory and/or field experiments.

3. Rainsplash detachment and overland flow

Two types of water erosion process operate on slopes. The first, rainsplash detachment, is caused by the impact of raindrops on the soil surface.

The second, overland flow or sheetwash, is caused by the rainfall forming a thin sheet of water flowing over the ground surface before joining up to produce small channels (rills) or much larger erosional features (gullies).

Although rainsplash is important for dislodging particles, the amount of erosion caused by this process is usually quite small because individual soil particles can only be moved short distances (10–20 cm). It is the combination of both rainsplash and overland flow operating on soils of different resistance to erosion and slopes of differing angle which controls the amount of erosion taking place. The following section describes a carefully designed experiment to test the effect of slope angle and soil type on splash and overland flow erosion.

3.1 Experimental design

In order to measure the effect of rainsplash, a number of splash trays of between 50 and 100 cm in width and 3–4 cm deep should be constructed as shown in Fig. 24a). These can be constructed from plywood, or plastic seed trays supported on a wooden frame could also be used. Holes should be drilled in the base to

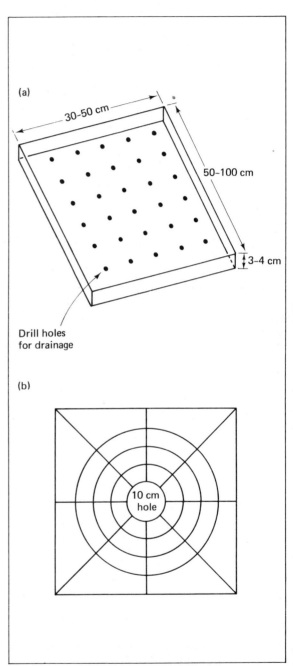

Fig. 24 An experiment to measure splash erosion
a) Design of the splash tray
b) Division of blotting paper into sections for splash erosion measurements

allow drainage, and blotting paper or filter paper placed in the bottom to prevent soil loss. The trays should be filled with sand (different particle size grades may be obtained from a builders merchant) or dried and ground soil. (Some knowledge of the size distribution of soil particles is needed here, see chapter 20). The sand or soil should then be carefully levelled across the tray using a straight edge. For rainsplash experiments, a sheet of blotting paper with a 10 cm diameter hole in the centre should be divided into quadrants and placed over the tray as shown in Fig. 24b). Several trays, containing soils of different particle sizes and/or standing at different slope angles (e.g. every 5° from 0 to 30°) should be placed in an exposed site out of doors until a rainstorm occurs. Rainfall should be measured by one of the methods described in chapter 13.

After rainfall has occurred, the blotting paper should be removed carefully, and the maximum upslope and downslope distance splashed (in cm) of any soil particle should be recorded. The paper should then be cut up into the sections as marked on Fig. 24b). Each section should be dried (preferably in an oven set at 105°C, or over a radiator if this is not possible), and accurately weighed. The dry weight of the blotting paper per unit area should also be calculated and subtracted from the measured weight in order to obtain the weight of soil splashed in each section. For overland flow measurements, similar trays to those shown in Fig. 24a) may be used, but the experiment is performed without covering the tray with blotting paper. In this case, a smaller tray, drilled with holes and lined with blotting paper is placed at the lower end of the soil tray to collect any sediment moving across the boundary. The blotting paper is dried and the total weight of sediment moved is also determined.

3.2 Analysis

The experiments outlined above can be used to demonstrate the importance of slope angle and particle size characteristics on the splash erosion and overland flow erosion process. The following analysis could be performed.

1. Use a system of shading and draw a diagram of the splash tray to show where different amounts of soil were caught.
2. Calculate the total, upslope and downslope weight of soil splashed. Plot each value against slope angle on a graph.
3. Plot maximum upslope and downslope distance splashed of any soil particle against slope angle on a graph.
4. Plot a graph of soil weight caught in overland flow troughs against slope angle.
5. Where several repeat experiments have been performed:—
    i) Plot the total weight of soil splashed against rainfall volume.
    ii) Plot the weight of soil transported by overland flow against rainfall volume.
6. Repeat the above analyses for different particle sizes.

References

Finlayson, G and Statham I (1980) *Hillslope analysis* (Butterworths).
Selby, M.J (1982) *Hillslope materials and processes* (O.U.P.)

19 Human impact on river channels

1. Introduction

River channels adjust themselves (i.e. change their size and shape) in response to the amount of water they carry. In small drainage basins, discharges are low and the channels small. In large drainage basins, the river channels become larger because they are carrying more water. Many studies have shown that as the size of the drainage basin increases so does the size of the river channel and that the relationship is approximately linear when the data are plotted on graph paper with logarithmic rather than arithmetic coordinates (Fig. 25). Within areas of similar climate, this broad trend will be fairly consistent unless other conditions increase or decrease the amount of water available within the river system. Long term climatic changes, for example, are known to affect the size of river channels because the relative balance between erosion and deposition will change as a result of the amount of rainfall which becomes runoff.

Channels are assumed to be in equilibrium with present flow levels; i.e. a balance between water discharge, sediment transport, erosion and deposition such that even where channels move across their floodplain they tend to maintain a fairly constant size. This equilibrium state may change as a result of human activity; such as that following replacement of natural vegetation by cultivated land or the process of urbanisation and reservoir construction. Urban development, for example, tends to increase the size (magnitude) of flood events for a given rainfall for a number of reasons.

i) Evapotranspirational losses become smaller because of the removal of vegetation.
ii) Infiltration rates are reduced because of the presence of impervious surfaces (buildings, roads, and pavements).
iii) Soil moisture and groundwater levels often decline because of reduced infiltration rates.
iv) Surface runoff increases, again because of the increase in impervious surface area.
v) Water arrives more rapidly in the stream channel because of the presence of gutters and storm drains designed to remove water from the urban environment more efficiently.

In the case of urban areas, the effect of increased flood magnitudes will decrease downstream as other unaffected tributaries join the affected one. Eventually the channel will take on characteristics indistinguishable from its natural condition. The amount by which channels have been modified can be measured in the field by looking at channel properties upstream and downstream of the affected area.

2. Aims

i) To carry out surveys of river channels in order to measure cross-sectional characteristics and particle size properties.
ii) To examine the impact of urban areas on channel and sediment characteristics.

3. Selection of drainage basin and survey sites

Catchments used for this analysis should be chosen with some care, for example large rivers with small villages comprising a few houses are unlikely to reveal any significant changes in channel form. Reference should also be made to geological maps to ensure that changes in channel or sediment properties downstream of urban areas are not simply a result of changing geology. Optimum basins are those usually less than $50-75$ km^2 in area, since larger rivers pose serious hazards during measurement. Catchments smaller than $10-12$ km^2 may be too small to detect significant changes in characteristics.

Specific locations for the measurement of channels should be selected in straight reaches of a river free of weed growth and overhanging roots. Since gravel bed rivers are characterised by alternating deep and shallow sections (pools and riffles), it is recommended that *all* measurements should be taken on the top of riffle sections for consistency. Exact selection of measuring points will depend upon local conditions, but all channel sections which have been canalised in concrete, or dredged, should be avoided. Approximate locations should be identified on a 1:25000 Ordnance Survey map, to provide measurements on approximately 8–10 sections above the urban area and should include all major tributary streams in addition to head water areas (Fig. 25b)). 5–10 equally spaced measurements below the urban area should be made on the main stream channel.

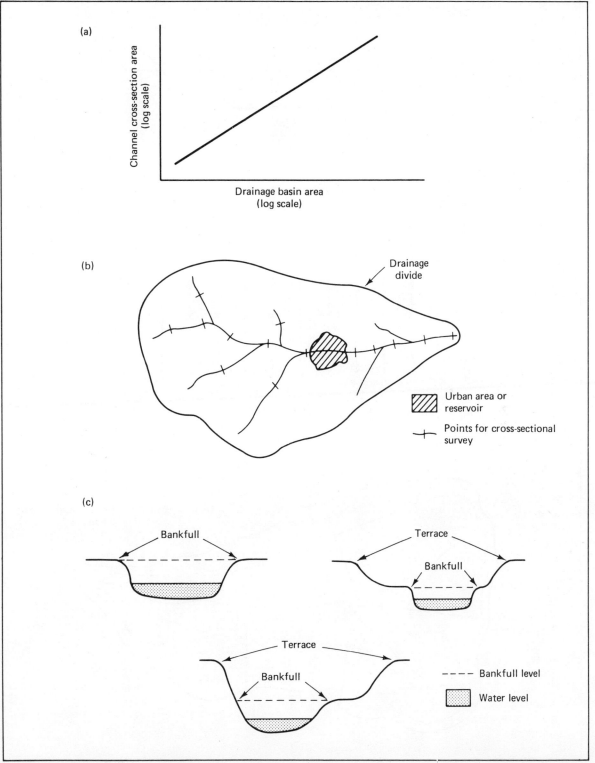

Fig. 25 a) The relationship between drainage basin area (log scale) and channel cross-sectional area (log scale)
 b) Location of measuring sections in a drainage basin
 c) Identification of bankfull channel capacity in the field

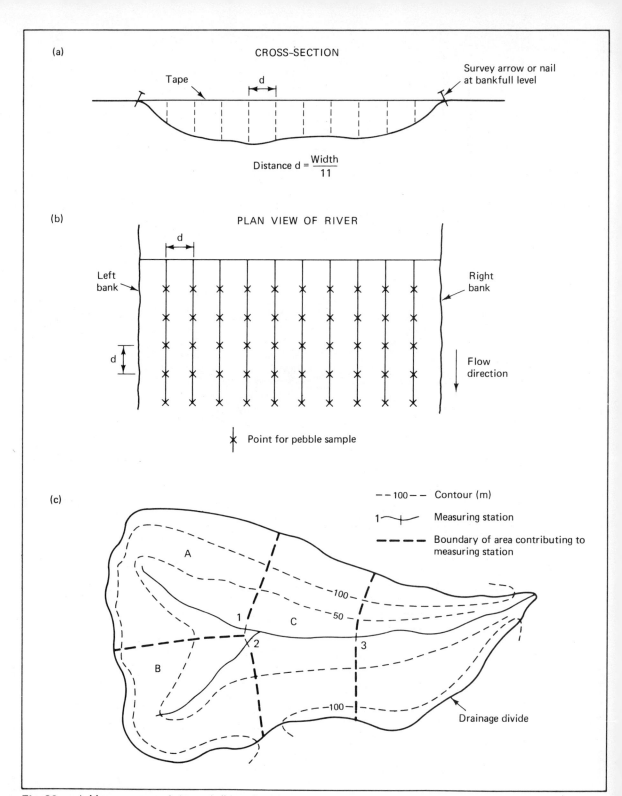

Fig. 26 a) Measurement of channel dimensions at a river section
b) Sampling scheme for particle size analysis
c) Calculation of drainage basin area above measuring sections

4. Field measurement

i) Locate a straight clean section of river, where the bankfull level of the river may be easily identified (Fig. 25c)). For most purposes the bankfull level is assumed to be the flow in the river which occurs for approximately 1 day every 1.5 years. It is thought to represent the same flow frequency for most rivers in the UK and is used to standardise the measurement of the channel between cross-sections irrespective of water level in the river at the time of measurement. It can usually be located in the field as the lowest break of slope in the channel cross-section. The absence of vegetation below this point can also be used as a guide to its location.

ii) In a river with pool and riffle sequences, locate the mid point of the riffle.

iii) Stretch a measuring tape across the stream and fix it around survey arrows (6" nails or wooden stakes will be suitable) as shown in Fig. 26a).

iv) Measure the bankfull width of the stream, and divide the width into 10–15 equal intervals.

v) At each interval use a metre rule or steel tape to measure the depth of the channel, and record the results as shown in Table 20. This will enable you to construct an accurate cross-section on graph paper in order to calculate bankfull cross-sectional area of the channel.

vi) Bed material size contained within the channel may also be analysed and details of the techniques available for this analysis are given in chapter 20. Such analysis will enable you to compare sediment characteristics above and below the urban area. Samples should be collected using a systematic random procedure such as that shown in Fig. 26b). At each of 50 points, a single sample should be randomly selected for measurement.

Table 20 Field Recording of channel cross-sectional data

Site Reference		
Distance (cm)	Depth (cm)	Notes
0	0.0	Bankfull level
10	8.0	
20	12.0	
30	14.1	
40	15.3	
50	14.4	
60	13.7	
70	12.6	
80	13.2	
90	11.0	
100	9.1	
110	0.0	Bankfull level

5. Laboratory analysis

From the tabulated results (e.g. as shown in Table 20) construct a scaled cross-section of the river channel on graph paper using the same vertical and horizontal scale. For example in small channels, 1 cm on the cross-section may represent 10 cm in the channel. If this scale is chosen, a 1 cm $\times$ 1 cm square represents 100 cm^2 of channel area, and a 1 mm $\times$ 1 mm square represents 1.0 cm^2 of channel area. Calculate the total bankfull cross-sectional area (in square metres) of each channel section. Calculate the mean depth of the stream by dividing cross-sectional area (in square metres) by channel width (in metres).

e.g. Cross-sectional area = 1.357 m^2
     Bankfull width = 1.872 m
     Mean depth = 0.725 m

For each cross-section located on the 1:25000 map, calculate the drainage basin area (in square kilometres: km^2) above that point. An example is given in Fig. 26c). The area contributing to channel section 1 is represented by area A and station 2 by area B. (The areas should be identified on the map with reference to the major topographic divides between basins). The area contributing to station 3 is represented by A+B+C. These areas may be calculated using a planimeter or by laying a square grid over the drainage basin and counting the squares.

Tabulate all data in order of increasing drainage basin areas as shown below:

Site no.	Grid ref.	Basin area km^2	Channel area m^2	Width m	Mean depth m
1					
2					
.					
n					

Draw three graphs to show the relationship between width, depth and cross-sectional area of the channel plotted against drainage basin area. These graphs should be drawn on graph paper with logarithmic axes. (Fig. 27). Draw the best fit straight line by eye through the points upsteam of the urban area and extend this line (extrapolate) in a downstream direction. (A best fit line using Pearsons product moment correlation and regression techniques could be used to good effect here, but remember first to take the logarithm of all values for the calculation and secondly to use only the upstream data for the calculation).

If the downstream values lie above or below the extrapolated line, this indicates that the channel is larger or smaller than would be expected. The degree to which these values differ from the expected may

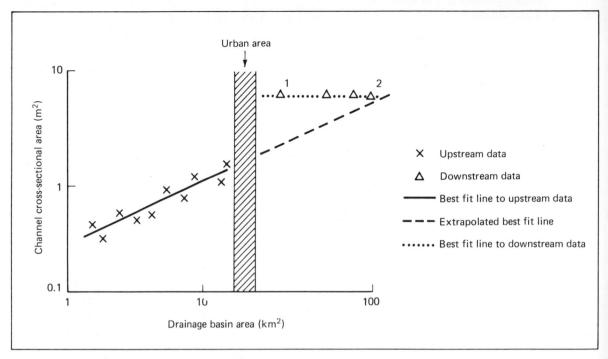

Fig. 27 Analysis of basin area and channel cross-sectional area data

be assessed by calculating the *'channel enlargement ratio'*. For example, the point labelled 1 on Fig. 27 has a cross-sectional area of 4.6 m². According to the extrapolated line, this should have an area of 1.4 m². The channel enlargement ratio is calculated from:—

$$\frac{\text{observed size}}{\text{expected size}} = \frac{4.6}{1.40} = 3.29$$

which shows that the area is larger than would be expected. For the point labelled 1 on Fig. 27 the observed value is 4.3 m² and the expected value is 4.2 and the channel is only slightly larger than expected. This indicates that the effect of the urban area decreases in a downstream direction as more natural tributaries contribute to flow.

The examples outlined above demonstrate the analysis of river channel cross-section data, but similar relationships could be plotted for stream width, stream depth, and the particle size distribution of the sediment.

Section 4 described a potential extension to the basic field survey by including information on particle size analyses. Where this theme is to be pursued, reference should be made to chapter 20 where simple graphical and statistical techniques are described.

References

Gregory, K.J and Walling, D.E (1976) *Drainage Basin Form and Process* (Edward Arnold).

20 Analysis of the physical properties of sediments

1. Introduction

One of the most important geographical processes is the transport of sediment through the landscape. Rocks are broken down by a variety of weathering agencies and are then transported by gravity or by wind, water and ice. Some of the sediment properties may change during transport, such as the rounding of pebbles by fluvial or marine processes. The sediment may be deposited at some point in the landscape and may remain there for any period of time from a few days to several millions of years.

Analysis of the physical properties of sediments may be used for two major purposes.

i) To study the nature of present day processes in order to examine changes in sediment characteristics through space or time. Within this context, it may be relevant to examine the direction and magnitude of changes in sedimentary properties and, on occasions, to examine the rate or speed at which changes take place. By using knowledge gained from these investigations it may be possible to illustrate the *nature* of processes and postulate the *results* of these processes in terms of landform development.

ii) To use sedimentary deposits as a historical record of processes acting in the past. For example, river terrace gravels or glacial outwash plains, will be indicative of the processes which led to their formation even though these processes may not have operated in the area for many thousands of years. These sedimentary deposits will provide information on the direction of transport and on the transporting environment. Furthermore, they may provide some indication of the climatic and broader environmental conditions at the time of deposition.

Of great importance in the analysis of sediments is the design of an objective and scientific investigation. This means that accurate *measurement* of the relevant sediment properties is essential. This is important because it may help to make observations and conclusions more precise when discussing how 'rounded' or 'angular' sedimentary particles may be.

2. Aims

1) To describe simple methods for determining the particle size and shape of sediments.

2) To describe graphical and statistical methods for comparing samples obtained from different environments.
3) To examine the application of these techniques in different geomorphological environments.

3. Determination of particle size

This is one of the most important and useful techniques for sediment analysis and, for large particles at least, presents few problems of measurement to the field worker. By convention, particles may be graded into a number of categories ranging from boulders which are larger than 256 mm in diameter to clays which are smaller than 0.002 mm in diameter (Table 21).

Since many different sizes of particle may exist in one area, no single measurement of a sample will provide a totally adequate description of the whole sample. For particles coarser than granules, however, some simple field techniques can be used as shown in the following sections.

Table 21 Classification of particle size

Type	mm Wentworth scale
Boulder	more than 256
Cobble	256 to 64
Pebble	64 to 4
Granule	4 to 2
Very coarse sand	2 to 1
Coarse sand	1 to 0.5
Medium sand	0.5 to 0.25
Fine sand	0.25 to 0.125
Very fine sand	0.125 to 0.0625
Coarse silt	0.0625 to 0.0312
Medium silt	0.0312 to 0.0156
Fine silt	0.0156 to 0.0078
Very fine silt	0.0078 to .0039
Coarse caly	0.0039 to 0.00195
Medium clay	0.00195 to 0.00098

3.1 Measurement

Many sedimentary properties can be determined by the use of accurate vernier calipers, a pebbleometer or a ruler. The objective is simply to measure a number of axes in a set coordinate system. Three axes are usually measured:—

i) The longest dimension of the particle (a axis)
ii) The longest dimension of the particle at right angles to the a axis (b axis).
iii) The longest dimension of the particle at right

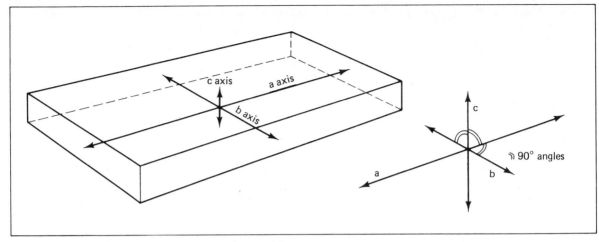

Fig. 28 Definition of the *a*, *b* and *c* axes for particle measurement

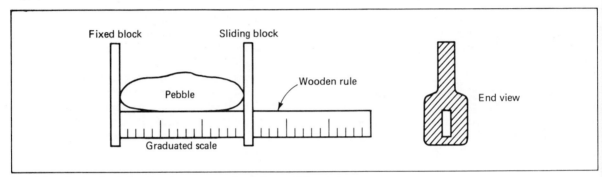

Fig. 29 A simple design for the construction of a pebbleometer

angles to the *a* and *b* axes (*c* axis).

An example of the measurement of the three axes is given in Fig. 28 for a flat tabular particle. The measurements in the field are best achieved by use of calipers or a pebbleometer. A simple design for the pebbleometer, using a wooden or metal rule with one fixed block and a moveable block, is shown in Fig. 29. As a last resort, a steel ruler or tape can be used to measure the three axes but this becomes difficult with irregularly shaped stones.

3.2 Procedure

i) Hold the particle in the pebbleometer to determine the maximum length (*a* axis): record the length.
ii) Rotate the particle through 90° so that the longest axis at right angles to *a* can be measured: record the *b* axis length.
iii) Rotate the particle through a further 90° and measure the longest axis at right angles to *a* and *b* as shown in Fig. 28: record the *c* axis length.

Note the values for each pebble in a field notebook using the following guide, and where appropriate, classify pebbles according to rock type, (e.g. flint, greywacke, granite, shale): a geological map of the area would be useful in this respect.

Location: — _____
Date: — _____ Time: — _____
Sample point on transect: — _____
Number of pebbles measured: — _____
Observer's Name: — _____

Sample No.	Rock Type	*a* axis mm	*b* axis mm	*c* axis mm	Size mm
1	Chert	63	47	20	43.3

The mean size of the particle is calculated from: —

$$\frac{a + b + c}{3} = \text{Mean size}$$

For the above example $(63+47+20)/3 = 43.3$ mm, according to Table 22 this particle may be classified as a pebble.

At any one sampling point, it is necessary to measure enough pebbles to provide a representative estimate of particle size for all rock types. No precise guidelines can be given, but as a rough guide a minimum of 30 stones should be measured if there is only 1 rock type, 60 if there are two rock types and so on. Do *not* specifically select a given number of pebbles of each type, but select each pebble randomly from a quadrat of between 0.25 and 1.0 m² in area (sides 0.5 m and 1.0 m respectively). The exact location of quadrats will depend on the objectives of the sampling exercise, and will be discussed after a consideration of the measurement of particle shape.

4. Particle shape

Although the size of a particle determines how much energy is needed to move it, other information can be obtained by measuring particle shape, because shape provides an insight into the means of transport. Angular particles, for example, are common in glacial environments and on scree slopes whereas transport in water leads to a rounding of particles.

Fig. 30 The Zingg classification of particle form showing the relationship to Krumbein's sphericity

4.1 Measurement

Precise measurement of shape is difficult because an attempt is being made to represent a three dimensional object by a single number. Consequently, several shape indices have been devised, but here emphasis will be placed on two simple measures, which may be calculated from a, b and c axis measurements described in the previous section.

4.1.1 Zingg's method

For each pebble, measure the a, b and c axes as defined above.

From these three measurements, calculate two ratios:

$$\frac{b}{a} \text{ and } \frac{c}{b}$$

From the previous example, $b/a = 0.75$ and $c/b = 0.43$.

Zingg classified particles into four sub-groups on the basis of the following conditions.

Classes	axes b/a	c/b
Spheres	>0.67	>0.67
Discs	>0.67	<0.67
Rods	<0.67	>0.67
Blades	<0.67	<0.67

In the example, $b/a > 0.67$ and $c/b < 0.67$ the particle may be classified as a disc. A simple graphical procedure can be used to produce a rapid classification after calculating the two ratios, by plotting the points of intersection in Fig. 30.

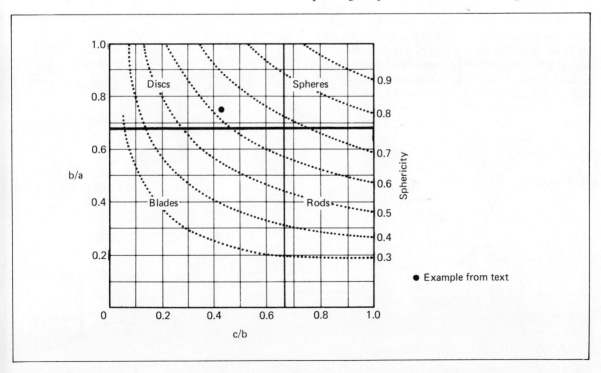

4.1.2 Krumbeins method

Krumbein proposed a measure which would relate the three measured axes (a, b and c) to the same measurements for a perfect sphere. It is calculated from:—

$$\text{Sphericity} = \sqrt[3]{\frac{bc}{a^2}}$$

In the example,

$$\text{Sphericity} = \sqrt[3]{\frac{47 \times 20}{63^2}} = \sqrt[3]{0.2368} = 0.6187$$

The sphericity value ranges from 0 to a maximum of 1 and a true sphere would have a value of 1. In reality, most sediments fall between sphericity values of 0.3 and 0.9. The relationship between the Zingg classification and Krumbeins sphericity index is also shown in Fig. 30.

5. Sample studies

5.1 Coastal environments

Two simple projects are particularly appropriate for the application of particle size and shape analysis.

The first attempts to examine the degree of particle sorting which occurs on different beach levels formed of shingle. These levels represent different 'energy' environments, with the highest level formed under storm conditions, and lower levels formed under lower energy conditions. Two specific hypotheses may be tested. The first suggests that only the largest particles are likely to be found on the highest level (nearest the cliff) since the smaller particles will be removed by storm waves and backwash. The second hypothesis suggests that pebbles on the highest storm beach level should be less rounded than those closer to the sea since the upper beach is less frequently affected by wave and tidal action. These two inferences are tested by setting up the approriate null and alternative statistical hypotheses and establishing a sampling programme which will allow objective acceptance or rejection of the null hypothesis.

Identify the number of beach levels present and mark out a transect from the cliff line to the low tide level as shown in Fig. 31. Use an inclinometer or slope pantometer to level the beach profile. Locate a 0.25 or 1.0m² quadrat in the mid point of each beach level (1, 2 and 3 on Fig. 31). Randomly select between 30 and 300 pebbles from each quadrat. The precise number will be conditioned by the overall objective of the exercise. A minimum of 30 is recommended, but if an attempt is also being made to compare the resistance of different rock types, then multiply the number of rock types by 30 and collect the same number of samples from each quadrat. Measure the size and shape properties outlined in the previous section and record the results in a field note book.

The second beach exercise relates to the degree of rounding and the reduction in particle size which may be associated with longshore drift. In this case, the transect should be set up parallel to the cliff. The distance between sample points will be conditioned by the length of the beach. Since particle size and shape will often change from the upper beach level to the lower level, the experiment should be carefully controlled by taking all samples from the same beach level or, where beach levels are not clearly defined, at some position midway between the cliff line and the sea.

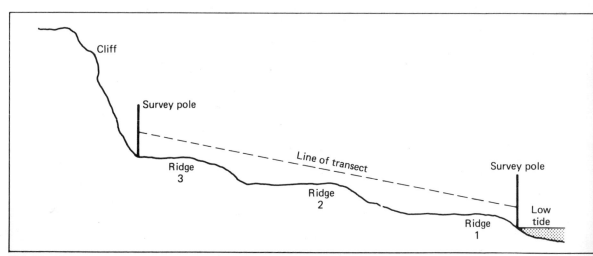

Fig. 31 Establishing a transect for beach pebble analysis

5.2 Fluvial environments

Particle shape and size analysis may be used to examine the changes which take place through the fluvial system in rivers transporting coarse bedloads. In this case, a number of sampling locations are established along a reach of river, preferably of several kilometres in length. Since rivers are characterised by alternating deep (pools) and shallow (riffle) sections, the sampling programme should be devised such that samples are always collected in the mid point of riffles (pools contain much finer sediments which can not be measured using the techniques introduced in this chapter). Again the sampling programme may be modified to examine the effects of rock type on the size and shape of particles and, where major tributaries join the river system, some attempt should be made to compare their properties with the main stream.

Many alternative experiments may be undertaken in fluvial environments, such as an examination of the changes in particle size and shape distributions across a single cross-section in meander bends and on straight reaches, or a comparison of the same properties in river terrace gravels and contemporary fluvial environments.

5.3 Analysis of slope deposits

Material found on scree slopes often moves only very short distances and cliff material which is removed by mechanical forces is structurally controlled and is usually angular. The material moving through such systems tends to be sorted by size but not by shape. Samples can be collected at points along transects laid out across scree slopes. Random samples are collected from quadrats located at equal distances along the transect down the line of maximum slope.

6. Presentation of results

Careful presentation of data collected from the sample exercises outlined above is important in order to produce a comprehensive report of results. The following examples show some of the methods which could be used.

Classification of particle shape using the Zingg method can be shown by plotting the data on Fig. 30. Alternatively, it may be represented by a simple histogram (or pie diagram) which enables comparison of particle shape according to rock type, as shown in Fig. 32a), or as a frequency distribution based on roundness classes as shown for samples collected from different environments in Fig. 32b).

Fig. 32 Graphical presentation of pebble analysis data
  a) Histogram showing the Zingg classification of particle shape sub-divided by rock type
  b) Roudness values for sediments derived from different sources

References

Briggs, D (1977) *Sediments* (Butterworths).
Finlayson, B and Statham, I (1980) *Hillslope analysis* (Butterworths).
Petts, G. E (1983) *Rivers* (Butterworths).

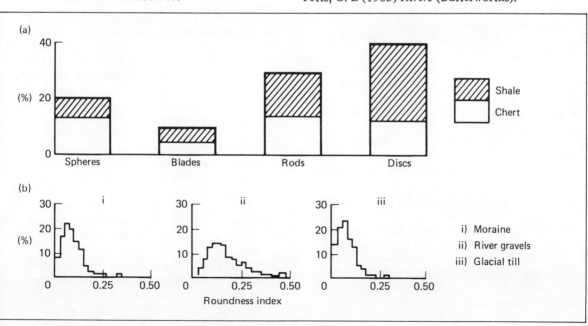

77

21 Describing and measuring soil properties in the field

1. Introduction

Accurate descriptions of basic soil properties are important for many types of biogeographical study. Any description provided, however, should be standardised with reference to techniques employed by the Soil Survey of England and Wales and, where possible, reference should be made to the Soil Survey Field Handbook compiled by J.M. Hodgson (1976). Individual surveyors will classify properties, such as colour, in different ways without a standard reference system. Compilation of soil survey information for the purposes of classification requires that all individuals describe the profile characteristics in the same way.

2. Aims

This chapter seeks to provide a number of simple techniques which will give a standardised procedure. The techniques are divided into three elements.
i) Site description
ii) Soil profile description
iii) Soil horizon description
Sample projects are discussed in section 6.

3. Site description

The site description should give general information with regard to three spatial scales, the regional scale, local scale and site specific scale. At the regional level, general information should include the following:—

Locality name	Date of survey
Ordnance Survey	Surveyors name
Grid reference	Weather conditions

In addition, a brief description of the regional relief and geology should be included.

The local relief, within an area of approximately 100 m of the profile site, should aim to identify:—
i) Precise location of site in relation to the region (e.g. interfluve, valley side, flood plain)
ii) Relationship to major drainage systems (sketch diagram)
iii) Aspect and local slope angle. The slope form (convex, straight or concave) should also be recorded.

The site features are those that immediately surround the point at which the soil pit is to be dug. Notes should be made on:—

i) Evidence of erosion and deposition (rills, gullies and mass movement)
ii) Evidence of flooding (trash lines)
iii) Land-use (by type of grassland or agriculture, tillage practice or type of forest)
iv) List of plant species and their dominance
v) Form of the ground surface (e.g. furrowed, stony or flat).

The above information can be recorded on a pre-prepared field sheet in order to catalogue the necessary information for the description of regional, local and site characteristics.

4. Description of the soil profile

Soil properties should be described in a rectangular pit dug to bedrock or to unconsolidated parent material and should be large enough to give the observer clear access to the lowest exposed face. Since the pit *must* be infilled after digging, a large polythene sheet should be laid adjacent to one side of the pit on which the excavated soil can be piled. This enables easy replacement of soil and turf and provides for minimum disturbance to the area.

The first task in describing the soil profile is to identify the presence of soil horizons. A horizon is a horizontal or near horizontal unit which differs in some way from the layers above and below. The distinction can be made on the basis of a number of characteristics such as colour, texture, stoniness or the presence of root systems. The horizon may be clearly defined or may grade slowly into the adjacent horizon and distinction should be made between the two when drawing a scaled soil profile description.

One of the most common systems of horizon classification divides the profile into three units, usually designated the A, B and C horizons:—

A This horizon is a mixture of mineral (parent material) and organic material. It is, in general, a horizon from which small particles, usually clays, are flushed downwards (translocated), and from which soluble salts are leached (the eluviated horizon). It is also the zone to which organic matter is added from the decomposition of leaf litter.

B This is a subsurface horizon where translocated clays and soluble salts are deposited after being removed from the A horizon. This is an illuvial horizon.

C   This is a mineral horizon derived from the weathering of parent material, but frequently retains some of the original structural characteristics of the bedrock.

This simple threefold classification may be adequate for some soils, but because there is usually a greater variety of profile types, further sub-divisions are usually needed. Where organic material is constantly being added to the soil surface, a threefold division is usually used.

L   The undecomposed Litter horizon comprising recently fallen leaves, twigs and branches.
F   The partly decomposed litter, still retaining part of its original form with clearly identifiable leaf structure: the Fermentation horizon.
H   The structureless fully decomposed Humus horizon containing only a trace of mineral matter.

The subsurface horizons may be divided into a number of subgroups, for example in the A horizon:
Ah is used to denote a dark humus rich horizon which is high in organic material.
Ap is used to denote a brown or red-brown soil horizon, low in organic matter and associated with ploughing in cultivated soils.

At the base of the A horizon, in some soil profiles, there is distinct evidence of the eluviation process which is seen as a very pale or light brown horizon. This is particularly common, for example, in podsolised profiles. It should be recorded in the following way:
E   an eluvial mineral horizon where clays are translocated and soluble salts removed.
Ea a bleached horizon, typical of podsolised soils (a = albic or ash coloured).
Eb a pale horizon (b = brown) which has a weak structure (crumbles easily in the fingers).

Within the B horizon, a number of subgroups may be identified by using the following suffixes.
Bh an illuvial humus horizon typical of podsolised profiles.
Bf illuvial chemical precipitate horizon, such as the iron pan in a podsolised profile.
Bt an illuvial clay horizon.

In addition to the nomenclature identified above, the following suffixes may be used for A, B or C horizons.
g   gleyed horizon (blue-grey colour) or mottled horizon caused by periodic waterlogging.
(g) slightly gleyed horizon.

When the soil profile is described, it is normal to measure the profile characteristics from the top of the A horizon, and each measurement is taken to the nearest 1 cm. Note should also be made in the scaled soil profile of the nature of the boundary between horizons: whether the boundary is sharp or whether

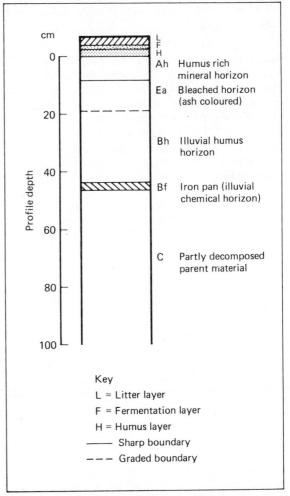

Fig. 33   The use of soil horizon nomenclature for a podsolised profile

there is a graded change from one horizon to another. An example of horizon nomenclature in use is given in Fig. 33 for a typical podsolised profile.

Once the soil profile has been classified on the basis of its constituent horizons, the next task is to provide a detailed description of the characteristics of each horizon.

5. Horizon description

The majority of the techniques for describing soil properties relate to the physical and chemical properties of each horizon. Many of the techniques employed by the Soil Survey require a detailed analysis of samples returned to the laboratory. This chapter concentrates on those characteristics which can be defined in the field.

5.1 Soil colour

Colour is difficult to define in an objective way without access to a standardised method of colour description. The Munsell soil colour system should be used, which contains precisely matched colour chips. The best known references are the Munsell Soil Colour Charts and the Fujihara Soil Colour Charts both of which employ the Munsell system of measurement. The system provides for a colour code based on a threefold division and also takes the form of a simple written description. It is based on the representation of colour in three ways; Hue, Value and Chroma.

i) Hue — This represents the dominant spectral colour and relates to the wavelength of light, (red, yellow–red, yellow etc.) and each colour is divided into four groups; 2.5 YR, 5 YR, 7.5 YR and 10 YR which are increasing hues in the yellow–red spectrum for example.

ii) Value — This is the apparent lightness or darkness of the soil and has a range of 1 (White) to 10 (Black).

iii) Chroma — This refers to the purity of the colour or the way in which the colour departs from white and neutral grey.

A small soil sample is moistened with water, because dry soils often show different colours from wet soils, and is compared with the printed colour chips in the Munsell colour chart to obtain as close a colour match as possible. The name and coded reference of the colour is then recorded for each horizon, e.g.

Hue	Value/Chroma	Description
10 YR	5/2	Greyish Brown
2.5 Y	4/4	Olive Brown
2.5 YR	3/2	Dusky Red

5.2 Soil texture

The texture of a soil is defined in relation to the amount of sand, silt and clay within the mineral portion of the soil sample, and may conveniently be divided into 12 separate categories (Fig. 34). The diagram is divided on the basis of the % of silt clay and sand and any individual sample can be classified after carrying out a detailed and time consuming particle size analysis of the sample in the laboratory. However, textural classifications can be made in the field by rubbing the sample between the thumb and forefinger, water is then added to the soil to observe its characteristics when saturated. The classification of the soil horizon into appropriate textural group can be achieved by reference to Table 22.

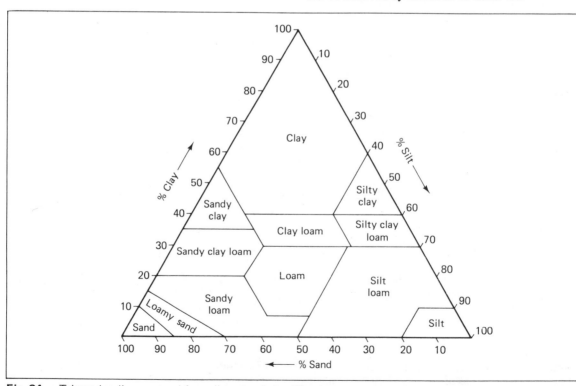

Fig. 34 Triangular diagram used for soil textural classification

Table 22 A field based textural classification of soils

Textural Class	Description of properties
Sand	Coarse — fine sand — visible grains loose when dry — not sticky when wet
Silt	Smooth soapy texture
Clay	Plastic and sticky when moist and can be rolled into thin threads. Takes finger-prints clearly
Silty-Clay	No sand — partly sticky when moist but has a smooth soapy feel of silt fraction
Silty-Clay-Loam	Trace of sand but enough silt to be slightly soapy. Less sticky than silty-clay
Clay-Loam	Sticky when moist — presence of sand can only be detected with great care
Loam	Moulds easily when moist and sticks to fingers. Can be moulded into threads but breaks easily on bending
Silt-Loam	Moderately plastic but not sticky — characterised by soapy feel of silt
Sandy-Clay-Loam	Slightly sticky when moist but sand fraction dominates
Sandy-Clay	Plastic and sticky when moist but sand fraction still obvious. Little soapy texture of silt
Sandy-Loam	Sand fraction dominates — moulds when moist — does not stick to fingers. Difficult to form threads
Loamy-Sand	Mostly sand — slightly plastic when moist — leaves smear on fingers when rubbed

5.3 Stoniness and mottling characteristics

The amount of mottling and the volume of stones present in the soil horizon should be estimated with reference to visual charts such as that given in Fig. 35 and should be described in the following way.

Stoniness %	Description	Stoniness %	Description
<1	Stoneless	20–50	Very stoney
1–5	Slightly stoney	50–75	Extremely stoney
5–20	Stoney	>75	Stone dominant

Mottling %	Description	Mottling %	Description
<2	Few	20–40	Many
2–20	Common	>40	Very many

Fig. 35 Chart for estimating percentage cover or composition

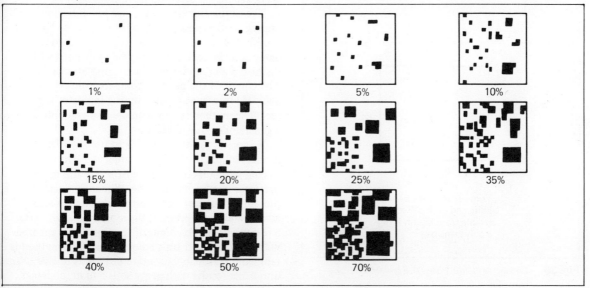

5.4 Bulk density

Bulk density is a measure of the mass of a soil per unit volume, in units of grams per cubic centimetre (g cm^{-3}). Measurement of this property requires the collection of a known volume of undisturbed soil. This is best achieved by using an open ended cylinder of approximately 5 cm diameter and 6 cm in length. (Large diameter central heating piping is very convenient for this).

The cylinder is pushed into the face of the appropriate soil horizon such that the end of the cylinder is flush with the face of the soil pit. It is then excavated with a trowel, and the other end trimmed flush with the cylinder using a sharp knife (Fig. 36). The sample should be transferred to a sealed polythene bag, because the sample may also be used for the determination of moisture content (chapter 22), as well as bulk density. Because bulk density varies with moisture content, the sample should be returned to the laboratory and oven dried (see chapter 22). Dry bulk density is calculated in the following way:

1) Volume of sample cylinder
   Internal Diameter   5.0 cm
   Length of Cylinder   6.0 cm
   Volume = $\pi r^2 h = \pi \times 2.5^2 \times 6.0 = 117.81$ cm^3
2) Weight of oven dried sample = 200.28 g
3) Dry Bulk density = $\dfrac{\text{dry weight}}{\text{volume}} = \dfrac{200.28}{117.81} = 1.70$ g cm^{-3}

5.5 Plants and animals

The most common method of quantifying the number of roots in each soil horizon is to mark out an area of 30 cm^2 on the profile face with a sharp knife, and then the number of roots in the area should be counted and classified in the following way:

Number of roots per 30 cm^2
>100   Abundant
 100–20   Frequent
  20– 4   Few
   3– 1   Rare

In addition to the quantitative analysis of soil roots, note should be made of soil fauna (such as earth worms) and their abundance.

5.6 Soil acidity

Soil acidity (pH) may be measured in a number of ways depending on equipment available.

The simplest method is to use pH indicator impregnated papers (e.g. BDH/Whatman or Hydrion types). A 3 cm strip of paper is pressed onto the damp soil surface, which may be moistened with distilled water, and is left for around 30 seconds. The colour of the paper is compared directly with standard reference colours on the indicator paper reel. The method is approximate and will only give a general idea of soil reaction.

A more accurate method is to use the BDH or similar soil pH testing kit. A small soil sample is shaken with a fixed quantity of water and universal indicator in a calibrated tube. Where fine particles are present, a few drops of barium sulphate are also added, and the mixture is thoroughly shaken. After a few minutes, the colour of the liquid is compared with standard tints on a colour chart. The method is rapid and reasonably accurate but tends to give unreliable results with peaty soils and in rendzinas.

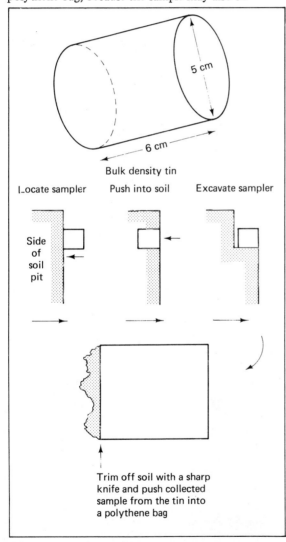

Fig. 36   Design and field use of a bulk density sampler

The most accurate method is to use an electronic field portable pH meter with a glass or plastic bodied pH electrode. The probe is calibrated in the field with reference to solutions of known pH. (These buffers can be made from tablets mixed with a known volume of water). Approximately 10 g of undried soil are mixed with 25 ml distilled or deionised water in a 50 ml beaker and left to stand for 10 minutes. The pH meter should first be calibrated against the buffer solution, rinsed in distilled water and immersed in the sample. The sample should be agitated carefully and the pH reading taken when the scale needle (or digital readout) is stable. The probe need not be buffered again for the next sample, but should be rinsed in distilled water.

6. Sample projects

Soil sampling and accurate soil profile description are important for many aspects of biogeographical study, including soil classification, and the testing of specific hypotheses relating to profile development. For this reason, many simple projects may be established which utilise the techniques described in the preceding sections. Three projects are briefly outlined in the following notes.

6.1 Soil catenary sequences

A soil catena (Latin=chain) suggests that similar soils develop in a sequence from a hill top to a slope foot position. Such developments relate principally to the amount of drainage because in flat areas, such as on ridge tops, drainage may be poor and soils may be peaty and/or gleyed. On valley sides, drainage is improved as slope angle increases and well mixed brown earth soils may develop. In valley bottoms drainage is again poor, and gleyed soils or peats would again be expected to develop. Drainage conditions in the local bedrock may serve to modify soil profile development, and the distinction between permeable and impermeable strata may give rise to local differences (see Fig. 37). These sequences of catenary development could be tested by careful description of profile characteristics at 4–6 places along a transect from a ridge top to a valley bottom in a geologically uniform area.

Fig. 37 Typical catenary sequences on impermeable and permeable bedrock

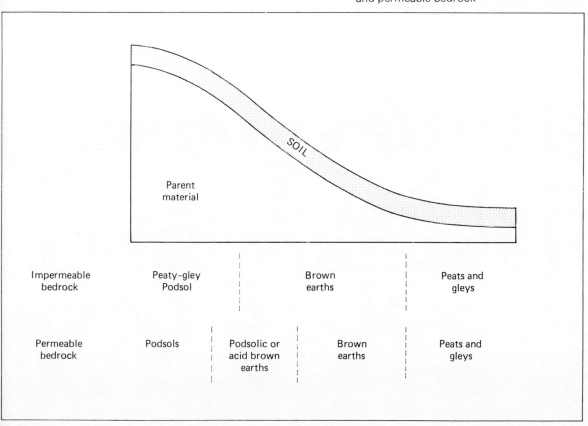

6.2 The effect of geology/parent material on profile development

Parent material is one of the most important factors controlling soil profile development. For example thin, nutrient poor, immature soils frequently develop on chalk and limestone (rendzinas) whereas deep, nutrient rich, mature well mixed profiles frequently develop on sandstones and marls (brown earths). The impact of parent material on soil profile development could be examined in a geologically diverse area, by describing and comparing profile characteristics. It is important, however, that all samples should be collected in the same relative position (ridge top, valley side etc.) because profile characteristics may change in a catenary sequence (see section 6.1).

Many soils do not develop directly on bedrock but on other superficial deposits (e.g. boulder clay, morainic deposits on soliflucted slope deposits). Suitable projects examining the properties and characteristics of soil profile development on different superficial deposits could also be designed.

6.3 The effect of vegetation and land use

Tremendous scope exists for the examination of soil profile development in relation to land-use and vegetation, and a variety of comparative projects may be carried out. For example, many forested regions contain a range of conifer and broad leaved tree species. Description and comparison of soil profile characteristics in relation to tree species could form an interesting and rewarding project. Alternatively, in cultivated regions, comparison could be made between soil profile characteristics in arable and grassland areas or in field systems in upland regions which have been abandoned in the last few decades or centuries.

References

Courtney, F.M and Trudgill, S.T (1976) *The Soil* (Edward Arnold).
Hodgson, J.M (1976) *Soil survey field handbook* Soil survey of England and Wales. (H.M.S.O.)
Trudgill, S.T (1977) *Soil and vegetation systems* (O.U.P.)

22 Analysis of soil moisture content

1. Introduction

Moisture content of a soil profile is important for a number of reasons since moisture content, along with the forces which hold water in the soil profile, determines the amount of water which is available to plants and also affects the way in which certain types of soil profile develop. The Podsol, for example, develops as a result of the downward movement or leaching of clay sized particles and oxides of iron and aluminium. This gives rise to the development of a grey or bleached A2 horizon and a red B1 horizon which is the zone of deposition. The factors which control moisture content in small areas are very complicated, but by means of a carefully designed experiment, the effects of land-use, soil and vegetation type on soil moisture content, may be investigated.

2. Aims

The aims of this chapter are
i) To describe methods for measuring soil moisture content.
ii) To describe a number of field projects to which these measures may be applied.

3. Field survey

Design of a suitable sampling framework is important for soil moisture measurement. Specific guidelines on the number of soil samples collected in an area is difficult to give, because soil moisture content may vary greatly over short distances. An experiment to identify micro-scale variations in these characteristics could be designed specifically to look at this problem (see section 5).

Sample locations should be selected on the basis of land use or vegetation type and, within the region chosen, a 10 m × 10 m grid laid out. Within this grid specific sites for soil sampling are selected by using random number tables. In general, at least 10 points for soil sampling should be identified on the basis of random coordinates, and at one location a soil pit should be dug and the profile described (see chapter 21).

4. The determination of soil moisture content

4.1 Field sampling

Soil sampling can be undertaken with a core sampler or screw auger where available. Alternatively, a soil pit can be dug with a spade and samples collected from each soil horizon with a trowel. Where average moisture contents are required for entire soil profiles, a pit of up to 50 cm depth (depending on profile thickness) should be excavated and samples of soil collected from each profile horizon. Where profiles show little evidence of horizon development, samples should be taken at regular increments of depth (approximately 1 sample every 5 cm) from the soil surface. Because soil moisture content will change with time, meaningful comparison of the moisture content of soils can only be undertaken if all samples from different areas are collected *on the same day*.

Once collected, samples should be wrapped in aluminium foil or in sealed polythene bags to prevent moisture loss and should be clearly labelled in waterproof ink noting site location and sample number. Long term storage should be avoided and all samples *must* be weighed within 24–48 hours of collection.

4.2 Laboratory measurement

Many methods exist for the measurement of soil moisture content, and the techniques outlined here are those which require the minimum of laboratory facilities.

Results from soil moisture calculations are usually expressed as a percentage of the weight of the dry soil. Table 23 shows the weight of a fresh soil, the weight of a dried soil and the weight loss after driving off moisture. Moisture content by weight is calculated from:—

$$\frac{\text{Weight Loss}}{\text{Dry weight}} \times \frac{100}{1} = \frac{2.89}{15.74} \times \frac{100}{1} = 18.4\%$$

4.2.1 Moisture content by oven drying

i) Take 15–30 g of soil and place it on a labelled foil tray of known weight. Weigh the sample and calculate wet soil weight:—
wet soil weight =
(weight of soil and tray) − tray weight

ii) Place the sample in an oven set at 105°C and leave it overnight to dry.
iii) Remove the sample to a desiccator (or sealed container with silica gel) for 1 hour. This allows the sample to cool without absorbing moisture from the atmosphere.
iv) Re-weigh the container and soil and calculate the dry weight:—
Dry soil weight =
(weight of dry soil and tray) — tray weight
v) Calculate weight loss from:—
Weight loss =
wet weight — dry weight
vi) Tabulate results as shown in Table 23 and calculate moisture content as explained above.

Table 23 Analysis of a soil sample for moisture content

Fresh soil weight (g)	Dry soil weight (g)
18.63	15.74
Weight loss (g)	
2.89	

4.3 Analysis of results

Moisture contents expressed as a percentage of either dry weight or as volume should be tabulated. Where details of moisture content with depth have been recorded, these should be plotted against soil depth, as shown in Fig. 38. Fig. 38a) for example illustrates a typical moisture content profile after a prolonged period without rainfall whereas Fig. 38b) illustrates a typical profile immediately after rainfall with the upper soil horizons wetter than the lower horizons.

Average moisture contents for each of the ten profiles in each grid may be calculated by summing the values for the profile and dividing by the number of samples. Comparison should be made between areas of different land-use, soils or vegetation.

Other measures of soil moisture, describing the spread of the values, may be readily calculated from the average profile data. One of the most useful is the coefficient of variation. This is calculated from the mean and standard deviation of the ten values from each sample area.

Coefficient of variation (%) =

$$\frac{\text{standard deviation}}{\text{mean}} \times \frac{100}{1}$$

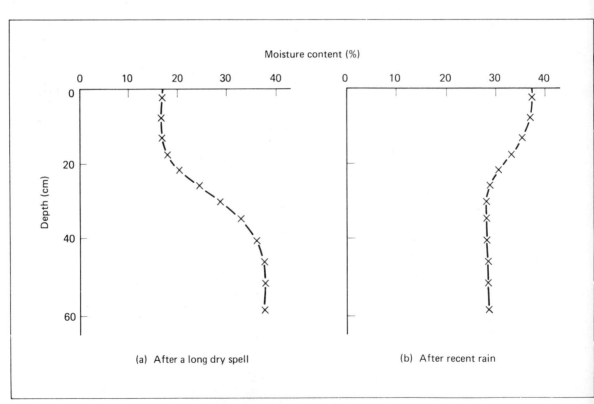

Fig. 38 Soil moisture content profiles for two soils

5. Sample projects

i) Comparison of soil moisture content stratified according to land-use, soil type or vegetation. In many areas for example, a variety of tree species (conifers and deciduous) are to be found. Comparison of properties between different species may provide an instructive field project in a small area. Alternatively similar measurements could be made in adjacent fields in cultivated regions containing grass, cereal and root crops since land management practices will also affect moisture content.

ii) The effects of sampling area size upon soil moisture properties could be undertaken by modifying the size of the sampling grid (for example in 5 grids with sides 1, 2, 5, 10 m in length) in the same location. Ten random samples would be taken from the grids of different size and the mean and coefficient of variation of moisture content would be plotted against grid size to examine the impact of sample area upon average moisture conditions.

iii) An examination of trends in moisture content along a hillslope transect from flood plain to interfluve would form the basis of a project examining the variablility in moisture content with respect to position in the landscape. Three to four sample grids are located on a transect running at right angles to the maximum valley side slope and at each site detailed soil profile description would be made in addition to some of the measurements detailed in this chapter.

References

Courtney, F.M and Trudgill, S.T (1976) *The soil* (Edward Arnold).

Knapp, B.J (1979) *Elements of Geographical hydrology* (Allen and Unwin).

23 An analysis of plant productivity and growth habit in bog communities

1. Introduction

Ecological studies form an important and integral part of biogeography and are based upon the well established ecosystem concept. Plants may be described in terms of their population characteristics (a term used to include groups of individuals of any kind of organism). Similarly the community includes all the population of a given area. The living community and the environment, in which plants live, function together as an ecological system; the ecosystem.

Although broad climatic controls affecting vegetation distributions can be identified at the world scale, micro-scale changes in the abiotic (non-living) part of the ecosystem may be equally important, not only to the type of species, population or community found but also to the growth rate and relative dominance of a particular species in a given environment. Even over distances of a few tens of centimetres, significant changes in community structure and growth habit may be observed.

2. Aims

The aims of this chapter are to
i) Examine methods which may be used to describe community structure and growth habit in a bog.
ii) Outline simple graphical and statistical techniques which may be used to analyse growth rates and habits in bog communities.

3. Description of bog communities

Bogs form one of the two subgroups of Mires; the second type is termed the Fen. Bogs are characterised by mosses (Sphagnum species) and dwarf shrubs whereas fens are dominated by monocotolydenous herbs, sedges, rushes and grasses. The two communities also differ in terms of pH, bogs usually being very acid, as low as pH 3.0 whereas fens are slightly alkaline (ph 7 to 7.5).

Within bog communities, there are usually microscale changes in topography, with mosses forming humocky terrain as shown in Fig. 39. These microscale changes in position and acidity may often influence plant community structure and development.

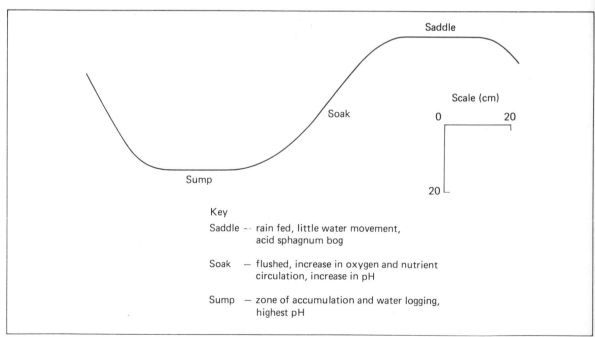

Fig. 39 Microscale variations in hummock-hollow topography in bog communities

3.1 Sampling and data collection

A study of bog communities at the microscale can be undertaken by setting up a transect of between 5–10 m in length across a selected area. The tape should be stretched in a horizontal position between two stakes located at the top of the hummock and a topographic survey undertaken by measuring the distance between the tape and the hummock surface as shown in Fig. 40. These data are used to construct an accurate scaled drawing of the bog surface and the transect is needed as a base on which to plot the distribution of plant species.

Individual plant species should be recorded along the line of the transect, providing presence and absence data only. This necessitates access to a good flora for species identification (see the books recommended at the end of this chapter).

Some species in bog communities occur in all positions, that is on saddles, soaks and sumps, but their rate of growth may be significantly different. Differences in growth rate may be calculated by measuring the length of approximately 50 growing leaves in random samples obtained from small quadrats (c 20 cm × 20 cm) subjectively located in all three positions in the bog community. Leaf length to the ground surface should be measured to the nearest mm with a ruler or metal tape. Species like the Bog Asphodel (*Narthecium ossifragum*) are particularly suitable because it produces long blade shaped leaves which may be easily measured.

At selected points across the transect (5 saddle, 5 soak and 5 sump positions as defined in Fig. 39) the pH of the soil should be measured as described in chapter 21). Calculate the mean pH value for each microenvironment. At the same points, measure soil temperature at 4 cm below the bog surface with a mercury thermometer and calculate the mean temperature for each microenvironment. Measure the shaded air temperature with a mercury thermometer at the vegetation surface in the same locations by the method described in chapter 13 and calculate mean temperature for each microenvironment. Where time permits, pH, soil and air temperatures can be recorded at 5–10 cm intervals along the transect after the flora have been recorded. It is important that temperature readings should be taken in as short a period of time as possible (no more than one hour) because significant diurnal changes occur over longer time periods.

3.2 Analysis

3.2.1 Presentation and tabulation of data

On the basis of the topographic survey, construct a scaled diagram to show the relief of the microcommunity. Plot the distribution of each species along the transect in separate diagrams. The distribution of each species should be examined in relation to tabulated mean values of pH and temperature or to similar transects of pH and temperate constructed from observations made at regular intervals along the transect.

3.2.2 Nearest neighbour analysis

An objective description of the distribution of species identified as discrete plants on the scaled diagram is provided by a modified version of the nearest neighbour statistic. This may be used to show whether the distribution is random, uniform or clustered. For example, if a single species occurs at seven points on a transect (Fig. 41) and are labelled 1–7. For each plant, establish its nearest neighbour (measured along transect) and tabulate the data in the following way.

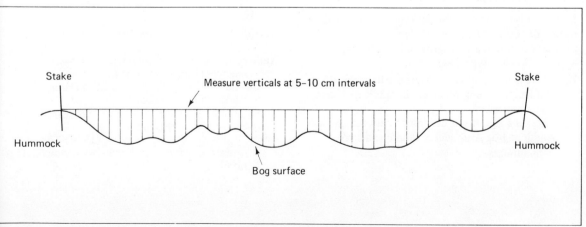

Fig. 40 A simple method for measuring micro-topography in bog communities

Plant	nearest neighbour	
1	2	} reflexive pair
2	1	
3	2	
4	5	
5	6	} reflexive pair
6	5	
7	6	

4. Additional comments

The techniques outlined above may be used for many microenvironmental investigations of growth habit and habitat, but are particularly suitable for bogs and marsh communities. It is important that great care is taken during sampling, with special regard to safety. Some bogs represent the accumulation of sphagnum over many hundreds or even thousands of years and are often several metres in depth. Even greater hazards are posed in floating bog communities where a carpet of sphagnum may overlie several metres of water (such as Llyn Mire in Radnorshire). Where in doubt about safety, local advice should be sought from the Nature Conservancy Council or local Park Rangers.

In two cases noted above, reflexive pairs are identified. This is where plants in two positions are reciprocal nearest neighbours. The nearest neighbour statistic is calculated by dividing the number of reflexive pairs (2 in the example) by the number of plants along the transect (7 in the example). The nearest neighbour statistic is therefore

$$\frac{2}{7} = 0.286$$

Test the significance of the nearest neighbour statistic in order to establish whether the distribution is random, regular or clustered.

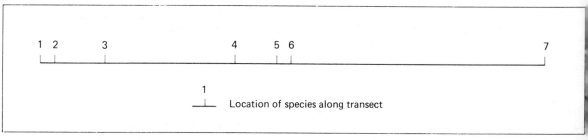

Fig. 41 Hypothetical distribution of a single species to demonstrate the calculation of the nearest neighbour statistics

3.2.3 Comparison of leaf lengths

Leaf lengths measured from soaks, sumps and saddles should be compared to see whether microenvironment affects growth habit by setting up the following hypotheses.

H_0 There is no significant difference between the leaf length of Bog Asphodel in saddle and soak communities.

H_1 There is a significant difference between the leaf length of Bog Asphodel in saddle and soak communities.

References

Watson, E.V (1963) *British Mosses and Liverworts* (C.U.P.)
Anderson, J.M (1981) *Ecology for environmental sciences* (Edward Arnold).
Odum, E.P (1975) *Ecology* (2nd Ed.). (Holt, Rinehard and Winston).
Keble–Martin W (1965) *The concise British Flora in colour* (Sphere) London.
Fitter, R, Fitter, A and Blaney, M (1974) *The wild flowers of Britain and Northern Europe* (Collins) London.

24 The analysis of vegetation associations

1. Introduction

Many factors will influence the type of vegetation growing in a particular area. Climatic factors, such as rainfall and temperature, are thought to control the distribution of vegetation in the world's major biomes, but other controlling influences are recognised in smaller regions. Of particular importance are the changes in soil type, which are partly a function of geology, and also the drainage of an area. The impact of fire and the destruction of natural forests, caused in part by human activity, has brought about dramatic changes in vegetation in all areas of the British Isles, including many upland regions such as Dartmoor, the Pennines and the North Yorkshire moors. These changes have produced conditions in which different types of plant community flourish today; ranging from alpine flora found in the Scottish Highlands, to heathland communities and semi natural deciduous forests in lowland areas. The type of vegetation which would be found in an undisturbed environment under given climatic conditions is referred to as climax vegetation, but in many areas such climaxes do not exist and sub-climax or subseral communities are to be found. In many cases, limits to development are caused by variations in relief, soils, drainage and a range of other controls such as slope instability features. These are referred to as 'arresting factors' because they prevent the development of a climax community.

2. Aims

The present chapter examines:
i) Objective methods for sampling vegetation communities;
ii) Some of the methods which may be used to describe plant communities in small areas;
iii) Simple methods for the graphical presentation and analysis of collected data.

3. Selection of sampling frameworks

The sampling framework should be selected objectively in order to minimise any bias in the sampling scheme that an observer may inadvertently produce. A variety of sampling strategies may be used, depending on the nature of the area to be sampled, and a discussion of the underlying principles of sampling design is given in Chapter 1. The types of sampling outlined below relate to the location of quadrats in which basic biogeographical and ecological information is recorded and a discussion of the methods which may be used to record this information is given in section 4.

3.1 Sampling procedures

Where a single vegetation community is sampled, a variety of random sampling methods are available. When a small area is to be sampled, the random points should be selected within an established grid framework. Although the Ordnance Survey grid is commonly used as a basis for selecting random samples from map sources, field sampling requires the establishment of a much smaller reference grid. The size of grid may vary considerably, depending on the number of samples to be taken, and the homogeneity of the plant community. As a rough guide, a grid of between 30 and 50 m side length would be adequate where between 15 and 30 sample quadrats are to be examined.

Procedure:-
i) Lay out a rectangular grid of side length between 30 and 50 m with a tape measure. (This can also be done with a nylon rope or string marked at 1 m intervals with coloured tape) as shown in Fig. 42a).
ii) Label one axis as a Northing (N) and one axis as an Easting (E).
iii) From a set of random number tables, locate the position of between 15 and 30 sample quadrats within the grid using numbers in pairs to establish the position of the Northing and Easting.
iv) Mark the location of the inter-section of each pair of random numbers. This forms the bottom left-hand corner of the sample quadrats which should be orientated within the coordinates of the main sampling grid.

The establishment of a random sampling grid is time consuming in the field, and an alternative method is to use a regular grid as shown in Fig. 42b), where the points of interaction of the grid mark the sampling positions. This method is based on the assumption that the vegetation is randomly distributed within the sampling area, and therefore that the sample is also random.

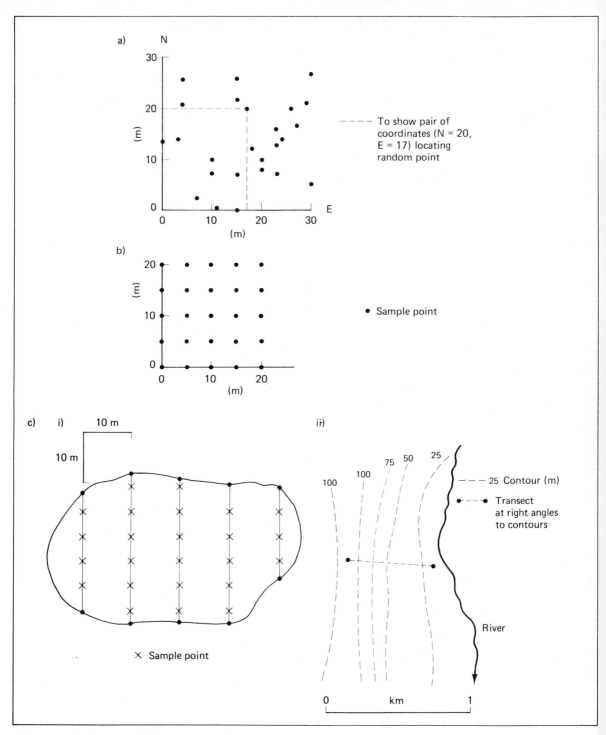

Fig. 42 Sampling strategies
    a) Random sampling in a rectangular grid
    b) Grid intersection sampling
    c) Transect sampling for i) Features of irregular shape ii) Valley side sampling

Both regular and random sampling techniques may be stratified if more than one vegetation community is to be described. This means that the community, and therefore the sampling areas, are selected by the field worker and are identified subjectively in the field. In many cases, the distinction between the two communities (such as a wet heathland and a dry heathland) is relatively straightforward.

In some cases, neither random nor regular sampling frameworks are viable methods of locating quadrats. This may happen when an attempt is made to examine the effects of other controlling variables, such as topography or position in the landscape, or when the area to be sampled is an irregular or linear shape, such as on drumlins or moraines formed in previously glaciated periods. In these cases, vegetation transects can be used to locate the position of sample points as shown in Fig. 42c). In both examples, transects are established across the feature of interest, and quadrats are located at equal distance along the transect.

4. Recording vegetation data in quadrats

A quadrat is a small area (between 0.25 and $1m^2$), in which information on species type, frequency and ground cover is recorded. The most important prerequisite for successful projects is acccess to a well illustrated flora to enable species identification. The works by W. Keble-Martin (1965) *The Concise British Flora in Colour* or by R. Fitter, A. Fitter and M. Blaney (1974) *The wild flowers of Britain and Northern Europe* are recommended. While most species are readily identified when in flower, the problem of of identification becomes more difficult in Autumn and Winter. This is especially true of the grasses (graminae) where many species can only be distinguished when in flower. Where the dominant ground cover species are not grasses, then projects may be simplified considerably by identification at the genus rather than the species level, or simply by identifying graminae as a single vegetation group.

Quadrats may be constructed simply and cheaply using a square wooden frame, nails and string as shown in Fig. 43a) which has 25 separate cells of 10 × 10 cm in length. The quadrat is located with the bottom left-hand corner at the randomly located point in the sample grid or at a point located on a transect. The data on plant characteristics, outlined below, should be recorded with the quadrat located at 4 separate positions (1, 2, 3, 4 in Fig. 43b)) to give a sampled area of 1 m^2 and 100 separate cells.

One of the most well established methods of recording plant community data is the Braun–Blanquet system which attempts to establish

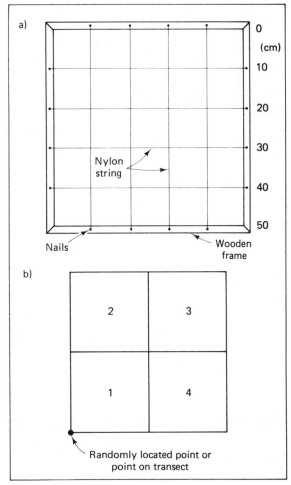

Fig. 43   Design and use of a 0.25 m^2 quadrat
a) Basic design
b) Use of 0.25 m^2 quadrat to give a sampling area of 1.0 m^2

sociological units within the vegetation. Description is semi-quantitative and, within each quadrat, all plant species are listed and their characteristics identified in the following way:

i) the frequency of each species is a measure of whether it is present or absent in each of the 100 sub-divisions or cells of the quadrat;

ii) the total area of the quadrat which the plant occupies, as a percentage of the total area, is the ground cover. It is an estimated measure (by eye) of how much light would be prevented from reaching the ground surface if the light source were directly overhead. Since vegetation is often layered, a total cover for all species may exceed 100%;

iii) the total area occupied by bare ground (no vegetation cover) should also be recorded.

Table 24 Field recording of quadrat data for vegetation surveys

Quadrat Number	Species A Frequency % Cover			Species B Frequency % Cover			Bare Ground % Cover
1	70	/	73	34	/	28	5
2	68	/	64	46	/	36	10
3	72	/	80	50	/	15	10

Table 25 Frequency and percentage of plant species in Wet and Dry Heathland Communities (25 quadrats randomly located in each community)

		Calluna Vulgaris	Pteridium	Erica cinerea	Erica Tetralix	Molinia	Ulex	Betula	Total
Dry Heath	Frequency	564	67	12	218	63	0	16	940
	% Frequency	60.0	7.1	1.3	23.2	6.7	0.0	1.7	100.0
Dry Heath	Frequency	427	18	61	18	0	51	0	575
	% Frequency	73.4	3.1	10.6	3.1	0	8.9	0.0	100.0

The information may be tabulated as shown in Table 24.

In addition, the sociability of each species should be recorded on the basis of the following 5 categories:
i) Isolated, a single plant
ii) Grouped or tufted
iii) Patches or cushions
iv) Small colonies or carpets
v) Pure populations

5. Analysis

Since a large quantity of data is generated by the description of a relatively small number of quadrats, it is important that the analysis provides a summary of the most important characteristics of the distribution of plant populations.

5.1 Graphical analyses

Graphical analyses of populations may be undertaken with reference to species frequency data. As an example, the data in Table 25 shows the frequency of occurrence of seven plant species collected from 25 randomly located 50 X 50 cm quadrats located in each of a wet and dry heathland community in south-east England. These data may be converted into percentages of the total plant population communities for each type of heathland by dividing the population of each species by the total population.

For example, for Calluna vulgaris on dry heathland,
Frequency of species = 564
Population frequency = 940
Culluna vulgaris as a % of total population =

$$\frac{564}{940} \times \frac{100}{1} = 60.0\%$$

The comparative data may be presented in the form of a bar graph, as shown for the same data in Fig. 44a). These results show, for example, that in both communities, Calluna vulgaris is the dominant species. However, Ulex is to be found only in the wet heath and Molinia and Betula are only to be found in the dry heath. Although Erica cinerea and Erica tetralix are found in both, the former is more common in the wet heath and the latter is more common in the dry heath. Such analysis provides a broad indication of the environmental tolerances of particular species.

Analysis can be made more penetrating by using a second graphical method which shows species diversity and species richness by plotting the distribution of the numbers of each species in rank order. Two extreme forms of distribution are found in natural habitats. These can either be a small number of species, dominated by one or two of them, or a large number of species with similar frequencies of occurence. This is shown in a hypothetical example in Fig. 44b) where the X axis represents the species ranked by order of abundance and the Y axis represents the number of individual species of each type. Curve i in Fig. 44b) shows a vegetation distribution characterised by a small number of species with only a small number dominant whereas curve ii in Fig. 44b) shows a large number of species of slowly declining dominance. Fig. 44c) shows the data of Table 25 arranged in this way and incidates:
i) dry heath has a larger number of species than wet heath;
ii) dry heath has a greater population of individual species than wet heath;
iii) both heathland communities are dominated by a small number of species.

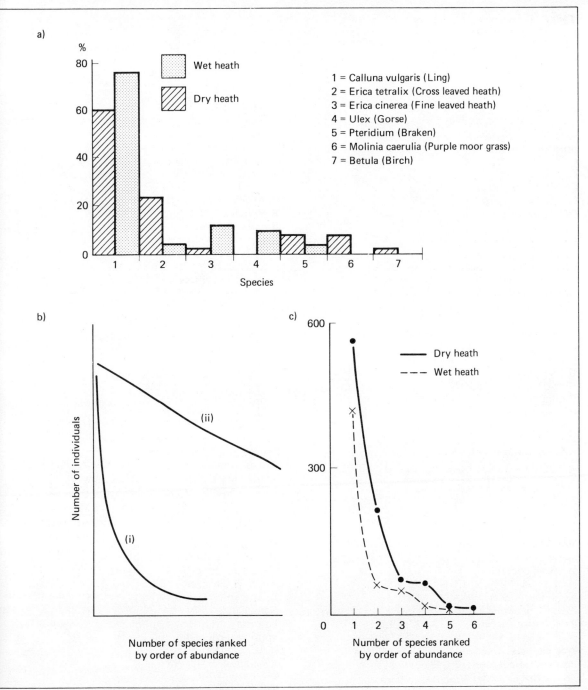

Fig. 44
a) Histogram to show species distributions in wet and dry heathland (Based on the data in Table 25)
b) Theoretical species diversity distribution curves (see text for explanation)
c) Species diversity distribution curves for the data of Table 25

References

Eyre, S.R. 1970 *Vegetation and soils.* (Edward Arnold).
Jones, R.L. 1980 *Biogeography* (Hulton).
Trudgill, S.T. 1977 *Soil and vegetation systems* (Oxford University Press).

Index

Accessibility index 9, 10
Agricultural intensity index 40, 41
Bankfull 68, 69
Bogs and mires 88–90
Catenary sequence 83
CBD 27–29
Census 9, 19–21, 38, 44
Central business index 27, 28
Centrality index 45
Chi-squared test 12, 33
Cloud cover and type 49
Complementary regions 44, 45
Components of industrial change 39
Core-frame concept 27, 29
Drainage basins 60, 68, 69
Environmental quality index 24, 25
Erosion 66, 67
Evaporation 48
Geomorphological mapping 61, 65
Hillslopes 66, 67
Housing quality 20, 21
Human impact 68–72
Humidity 48
Interception 51–54
Landscape evaluation 13–15
Land Use:
   agricultural 40, 41
   industrial 37–39
   rural 16–18
   urban 19–21, 24, 27–29
Location quotient 21
Lysimeter 55, 56
Mapping 61–65
Meteorology 47–50
Meteorological station 47
Moisture content 85–87
Nearest Neighbour analysis 46, 89, 90
Neighbourhoods 30–33
Parish summaries 40
Particle size and shape 73–77, 80, 81
pH 82, 83
Precipitation 48
Pressure (atmospheric) 48
Productivity 88–90
Quality of life index 23, 25
Questionnaire surveys: 6, 7
   consumer 11, 31, 45, 46
   farm 41, 42

   industrial 38, 39
   neighbourhood 30–33
   quality of life 23, 25, 26
   recreation 16–18
   retailing 35, 36
   rural accessibility 9–12
Rainfall 48
Rank correlation 15
Rate index 21
Retail environments 34, 35
Rivers: 57–60, 68–72
   channels 68–72
   discharge 57–60
   pools and riffles 71
   velocity 57, 58
Safety 8, 90
Sampling: 6, 7, 9, 13, 16, 20, 25, 27, 30, 34, 37, 40, 44, 52, 53, 70, 71, 76, 77, 83, 85, 89, 91–93
Secondary data sources:
   agricultural 41
   industrial 37, 39
   retail 34, 45
   rural 18
   urban 19–21, 27
Sediments 73–77
Service hierarchy 44–46
Service provision 9–11
Slopes 61–65
Social class 20
Socio-economic group 21
Soil: 55, 66, 67
   moisture 85–87
   profile description 78–84
Standard industrial classification 38
Standard labour requirements 41
Stevenson screen 48
Sunshine recorder 49
Temperature 48, 49
Throughfall 51–54
Tourism 16–18
Transpiration 48
Urban models 19
Vegetation 88–95
Visibility 49
Water balance 55, 56
Weather: see Meteorology
Wind 48